# GEOMODELS IN ENGINEERING GEOLOGY: AN INTRODUCTION

**Peter Fookes**
Consultant Engineering Geologist, Retired, Winchester, Hampshire, UK

**Geoff Pettifer**
Engineering and Environmental Geologist, Mitcham, Surrey, UK

**Tony Waltham**
Engineering Geologist and Karst Specialist, Nottingham, UK

Whittles Publishing

*Published by*
Whittles Publishing,
Dunbeath,
Caithness KW6 6EG,
Scotland, UK

www.whittlespublishing.com

© 2015 P. Fookes, G. Pettifer, A. Waltham
ISBN 978-1-84995-139-5

Printed by
Charlesworth Press, Wakefield

# Geomodels and pictorial block diagrams

*Geomodel is the general term for any form of interpretation of a geological situation. Several other relevant terms are used throughout the book to describe more specific geomodels.*

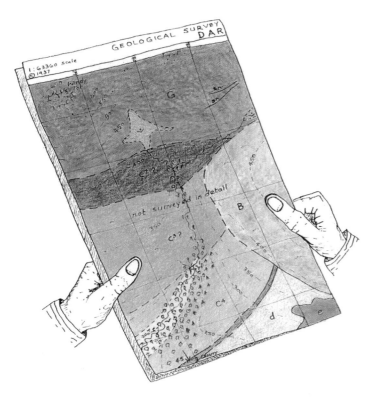

**The geological map: the vital starting point for any ground investigation**
**(see Part 4)**

The preparation of each pictorial block diagram in this book typically began with a conceptual discussion, followed by an examination of maps, the published literature, photographs and the development of a sketch model based on experience. This was then reviewed. The final diagrams were produced manually by placing a sheet of tracing paper on a template and, using a soft pencil, drawing a block outline from an arbitrary viewpoint to cover the area of interest and important locations. Approximate distances were estimated along the horizontal axes of the block, the topography was sketched using an appropriate vertical exaggeration and the surface drainage was added. The underlying stratigraphy and geological structure were then added, the probable groundwater conditions assessed and adjustments were made to highlight important landforms. Finally, the lines were inked in, the pencil work erased, the key features emphasized and labelled and captions added. Layers of colour were then applied by crayon on a good paper copy.

It is not suggested that this method is followed during a ground investigation to produce site geomodels, but the same procedures can be used to produce block diagrams for important engineering projects. These would also incorporate specific information from ground investigations, satellite images and site reconnaissance mapping. Site models (ground models) are primarily intended to aid site evaluation and engineering decision-making, but they can also be used to inform a wider non-specialist audience. Larger scale block geomodels covering small areas may be useful in illustrating how specific landscape features have evolved, or are expected to evolve, to update knowledge of ground conditions as more data become available, or to help solve specific problems arising during investigation and construction. The production of such diagrams is a time-consuming process. Software packages can now be used to produce the basic block outline and topography, but an interpretation of the 'total geology' generally has to be built up using freehand methods. It is therefore likely that for the time being 'traditional' field maps, sketches and cross-sections will continue to be the most practical form of geomodels for smaller projects.

# Contents

# Contents

# List of Tables

# Preface

For many years I have had occasional requests for permission to use, for teaching purposes, copies of the pictorial block geomodels and the related two-dimensional figures and tables originally published in the first Glossop Lecture of the Engineering Group of the Geological Society of London (Fookes, 1997a). I have therefore come to the conclusion that it would be helpful to publish, in a collected form, not only the original block models, but a few subsequent models that have been published elsewhere. Hence the idea for this book.

The aim of this book is to provide an introduction to geomodels in which the drawings and photographs largely speak for themselves. It aims to help engineers to visualize the three-dimensional geology and to act as a quick introduction to new or unfamiliar ground or environments for geologists and engineers. It could perhaps also be used as an aide mémoire to those who might be more familiar with the selected environments, and last, but not least, to be used for teaching. The pictorial models are the springboard for developing the geomodel approach as a tool for use in ground investigations and to help the development of geotechnical models for design and construction. The development of site geomodels (ground models) during investigation is therefore an important part of this book. The text is meant to be straightforward with few frills, i.e. the field engineers' or geologists' KISS principle: 'Keep It Simple, Stupid!' – I first heard of this a few decades ago when in the field in Western Australia and have overused it ever since.

Since their original publication as black-and-white models in 1997, colour has been added and this has significantly improved the originals. It must be emphasized strongly that these are very basic generic models; they are not comprehensive and no doubt contain many arguable points when viewed by a specialist in a particular terrain or environment. The authors do not claim such specialist knowledge other than that acquired in their careers as engineering geologists. Many good textbooks exist on the various specialist subject areas, written by those with better subject knowledge than ourselves.

The book is arranged in five parts dictated by the existence of the pictorial block models and the related two-dimensional figures and tables, to which one completely new block model has been added to give supporting detail.

*Part 1* introduces some necessary basic fundamental geology.

*Part 2* looks at common natural surface and near-surface conditions that have been modified by current and former climatic regimes at the location. Such modifications (e.g. tropical and temperate weathering, periglacial freeze–thaw activity) are typically not sufficiently well known to engineers who have not previously worked in these environments. The models in this Part are intended to help them in their investigation, design and construction activities.

*Part 3* delves further into the main geo-environments with the aim of helping to understand the engineering geology characteristics of a particular environment.

*Part 4* shows how knowledge of the ground improves during the various stages of a ground investigation with the help of site geomodels developed as the investigation progresses (perhaps the most important part of the book).

*Part 5* is loosely based on case histories to illustrate some minor to major pitfalls in investigation and construction situations. It was the most enjoyable part to write, but the most difficult for which to find photographs. Hence we have taken the opportunity to put in some small tables.

Geoff Pettifer originally drew all the figures and blocks for me professionally, including the colouring at a later date, with meticulous care and skill, typically taking more than one month to complete each drawing. I have long thought that his particular strength in producing this artwork has been insufficiently attributed in previous publications and hope that this book makes amends. Tony Waltham, as well as being an experienced, practical engineering geologist and consummate communicator and friend, is acclaimed for his field photography and has provided nearly all of the photographs in this book.

P.G. Fookes, F.R.Eng.
Winchester
April 2014

# Dedication

To our friends at Sunrise of Winchester (P. F.)

To Ann, our daughters and grandchildren (G. P.)

To Jan, who is the scale on so many of the photographs (T. W.)

and to the many geologists and engineers who have given us invaluable advice and support during field
and construction projects around the world

# The Authors

**Peter G. Fookes, F.R.Eng.** is a retired international consultant in engineering geology, geomaterials and concrete technology with well over fifty years of working professionally in some 96 countries. His work has been mainly in large, heavy civil engineering and open-cast mining, including bridges, dams, harbours, highways, pipelines, railways and tunnels, in deserts, mountains, permafrost and rainforests, in such diverse locations as Australia, Borneo, Brazil, The Falklands, Kenya, Libya, Nepal, Papua New Guinea, Siberia, Sudan and most places in between. He was/is a Visiting or Honorary Professor at several universities, has published over 200 refereed papers and articles, also seven books as editor or author, and has chaired or been a member of many national and international committees and working parties.

**Geoff Pettifer** is an active Chartered Geologist with over 35 years' experience, ranging from microscopic and X-ray examination of aggregates, site investigations, rock cutting design and earthworks supervision, to regional landslide studies and flyover terrain evaluations. He has postgraduate qualifications in Geomaterials and in Environmental and Earth Resources Management and has worked on projects in Europe, Africa, South America and the Asia Pacific region, including pipelines, power stations, open pit mines, quarries, railways and roads. As a member of multidisciplinary teams, he has contributed to ESIA reports and capacity-building manuals, and has participated in industry-based research projects dealing with chemical impurities in concretes, the effects of clay minerals in tunnelling and the capabilities of chain trenchers.

**Tony Waltham** is a retired lecturer in engineering geology, who primarily enjoyed teaching the subject to students of civil engineering, and developed his lecture notes into a textbook now widely used in universities. He also pursued research and consultancy in ground subsidence, particularly related to sinkholes and collapses in limestone karst, where a deeper understanding was aided by his many years of cave exploration. He has written and edited numerous books and academic papers. At the same time, his worldwide travels have enabled him to build his own extensive photograph library of geographical and geological subjects.

# Foreword

It has been said of Karl Terzaghi, the father of geotechnical engineering, that he turned every field trip into a joyous adventure in field geology. I have been on field trips with Peter Fookes and he does exactly that. For those not fortunate enough to have accompanied Peter on a field excursion this book is the closest that you can come to that joyous adventure.

A knowledge of the ground profile and its genesis together with groundwater conditions is arguably the most important body of information required for planning and designing a major geotechnical, quarrying or mining project. Unlike most structural engineers who usually work with well-defined geometries and specify their materials, geotechnical engineers have to work with materials as they have been laid down by nature – and nature is seldom simple. A structural engineer working on an ancient cathedral (perhaps to stabilise, conserve or extend it) has a number of challenges that are similar to those faced routinely by the geotechnical engineer. First it is necessary to discover how the cathedral was built over the centuries and the order in which the various elements were constructed. This may be thought of as discovering the genesis of the building and without this information it is not possible to carry out a reliable analysis of the 'flow' of forces through the various elements. Then it is necessary to determine the properties of the various elements and usually there is very limited scope for extracting samples and testing them – much reliance has to be placed on experience. A vital part of the investigation is to identify hidden weaknesses and defects and this requires detailed visual inspection and a trained eye – sometimes coupled with measurements of movements. Today it would be usual practice to carry out a detailed structural analysis of the cathedral and the proposed works.

In simple terms this is an exercise in 'structural modelling' which requires a knowledge of the genesis of the building, the material properties and the defects and weaknesses. Without this knowledge even the most sophisticated analysis would be a waste of time and could be very misleading.

Geotechnical engineers daily face similar challenges encountered by the structural engineer working on an ancient building or monument. If they are to model successfully the impacts of a proposed project it is vital that they understand the genesis of the geological formations they are to encounter, have a good knowledge of the mechanical properties of the materials and be aware of likely defects and weaknesses. This book focusses on understanding 'what is there, how it got there and when it got there'. It is about 'geomodelling' and its importance in all aspects of ground engineering from initial desk studies through site and ground investigation, testing, design, choice of construction method to construction and operation.

I mentioned that a field trip with Peter Fookes was a joyous adventure in field geology and I urge you to take a stroll through Part 1 – *Underlying factors: climate and geology*. Much of the material may be familiar but Geoff Pettifer's splendid block geomodels and Tony Waltham's magnificent photographs bring it to life in a new and fresh way.

I hope you have been hooked! Part 2 – *Near-surface ground changes* is a must for geotechnical engineers working on construction problems. It contains lavish illustrations of a wide range of geological and climatic processes that underlie the formation of soil profiles worldwide.

Part 3 – *Basic geological environments for engineering* gives excellent examples of the way in which the various past and present climatic environments have profoundly influenced both terrain and structural features of the ground. Many important engineering implications are illustrated emphasising the necessity of understanding the geological processes that form the ground profile.

Parts 4 and 5 move on to the engineering aspects of geomodels. Part 4 – *Ground investigations* describes the staged process of gaining knowledge about the site. Peck (1962) in an essay on *Art and science in subsurface engineering* (Géotechnique, 12:1, 60–66) asserts that 'whether we realise it or not, every interpretation of the results of a test boring and every interpolation between two borings is an exercise in geology. If carried out without regard to geological principles the results may be erroneous or even ridiculous'. This section gives graphic illustrations of how the staged evolution of a geomodel for a site assists in the selection of appropriate locations for borings and provides the geological context for interpreting the results of these boreholes. Case histories are vital to developing experience and judgement and Part 5 – *Case histories and some basic ground characteristics and properties*, in keeping with the earlier parts of the book, provides copious well-illustrated examples of the challenges of interpreting structurally complex geology.

For students studying engineering geology and geotechnics this book will provide an invaluable insight into the art of unravelling the complexities of the ground. Practitioners will wish to dip into it as they encounter a variety of ground conditions and terrains. Anyone interested in the way the landscape is formed, whether or not they are engineers, will find this book fascinating.

John Burland, CBE, FRS, FREng
Imperial College London

# Acknowledgements

I give thanks to the many friends and colleagues who, over five or six decades, have helped me, in the field or laboratory, to learn more and to clarify my understanding. I can name here only a few: first and foremost, Sir Alec Skempton (polymath) and Ian Higginbottom (engineering geologist); also, during all of my early and middle career, John Atkinson, Ken Head (soil laboratory), Alan Little, Mike Sweeney and Peter Vaughn (soil mechanicians); Lawrence Collis (materials engineer and concrete laboratory); latterly, Mark Lee (engineering geomorphologist), Fred Baynes and John Charman (engineering geologists); and, for shorter periods, but still with invaluable interaction, Jim Griffiths, Andrew Hart and Gareth Hearn (engineering geomorphologists), David Shilston (engineering geologist), Ian Sims (materials geologist and petrology laboratory), Jim Clarke (geotechnical engineer), Mike Walker (concrete and structural engineer) – all knowledgeable, clear thinkers who have on many occasions helped me to get my thoughts straight.

P.F.

I acknowledge the invaluable support of colleagues stretching back to my formative time at Wimpey Laboratories. Marjorie Eglinton was Chief Chemist and Ian Higginbottom headed a team of geologists that included Jasper Cook, Martin Dawes and Dave Earle. Since then I have had the privilege of working with and learning from many other engineering geologists and geomorphologists, in particular Fred Baynes, Andrew Hart, Gareth Hearn, Alemayehu Mulachew, Alan Poole and Mark Ruse. I also thank the numerous site engineers who gave me unstinting practical advice during construction site visits, among them Peter Bel-Ford, Solomon Kuliche, Mark Miyaoka, Neil O'Donnell, Ian Steele and Hugh Ward.

G.P.

For providing guidance over the years and exhibiting qualities that I respect, I thank Alistair Lumsden, Neil Dixon, Steve Hencher, Peter Smart and Art Palmer: from them and from others there is always much to learn.

T.W.

# Introduction

## WHAT, WHY AND WHEN?

The Earth is an active planet in a constant state of change. These changes can take place over both long and short periods of geological time (thousands or millions of years) or much more quickly on an engineering timescale (minutes, hours or days). Geological processes continually modify the Earth's surface, destroying old rocks, creating new ones and adding to the complexity of ground conditions: the so-called 'geological cycle'. The all-important concept that drives this geological cycle is plate tectonics (see Figure 1.3).

The benefits geologists bring to construction projects must exceed the cost of their services – that is, they must accurately improve the engineer's ground knowledge more cheaply and effectively than any other method. They must reduce the risk from geological hazards by anticipating situations perhaps unforeseen by the engineers and also help to determine effective ways of dealing with risks and any problems arising during design and construction. The main role of the engineering geologist is to interpret the geology and ground conditions correctly. Creating an initial model for the geology of a site is an excellent start. Geology (the study of the Earth) and its closest geo-relative, geomorphology (the study of the Earth's surface), are concerned with changes over time and any geomodel has to build in any changes likely to occur in the near future, especially when the construction project may have a significant impact on the environment (Fookes, 1997a).

Some degree of uncertainty will always exist in both the interpretation of the existing geology and any anticipation of significant changes over the lifetime of the project. The key types of uncertainty include data uncertainty and environmental uncertainty. Some aspects may defy the precise prediction of future conditions – for example, earthquakes, landslips, flooding, our limited understanding of the behaviour of complex Earth systems, or future choices by governments, businesses or individuals that will affect the socio-economic or physical environment. Many assessments rely on expert judgements based on the knowledge available, together with experience from other projects and sites. The problems associated with expert judgements include the poor quantification of uncertainty, poor problem definition and any bias of the assessor.

To paraphrase the above, many, if not most, difficulties in ground engineering arise from either an unawareness of the ground conditions or a failure to appreciate the influence of the known ground conditions on a particular engineering situation.

## GEOMODELS

### Geological pictorial models

The three-dimensional models in this book are not comprehensive, in particular those on river systems and landslides (Part 3), and should be considered as an introduction to landforms. They are mainly intended for teaching and are too simple for the more detailed prediction needed on an engineering site when a more detailed understanding of the near-surface geology is required. This is developed through the stages of a ground investigation and it is necessary to build a site geomodel for practical use (see Part 4). Such a model is based on engineering geology and ultimately leads to a 'geotechnical model'. We consider geotechnical models to be an important subject, but they are not discussed further in this book, except to emphasize that the engineering geology model broadly leads to a geotechnical model and other end-products.

Much has been published in the last decade or so on the use of models in engineering geology, building on a very small number of such papers before this time. It is not the intention of this book to enlarge upon the various approaches to, and philosophies of, building geomodels, the names given to these models and the uses to which they can be put. These depend on the actual project. For example, in the pipeline investigation industry, in which we have worked for many years, the term 'ground model' is being increasingly used. Specialist teams (geoteams) are developing in this industry. It must be emphasized repeatedly that, at any scale, every geomodel is different and must be tailor-made for the current project. There is no ideal model, just relevant approaches and good practices for the situation. Morgenstern and Cruden (1977), Fookes *et al.* (2000), Sweeney (2004), Baynes *et al.* (2005) and the many references cited in these papers provide useful background reading on models and model-making.

It should be noted that during the closing stages of preparing this book, a comprehensive study, *Engineering Geological Models: an Introduction* (IAEG Commission 25; Parry *et al.*, 2014), was published. This will be a most valuable addition to the field and office libraries of site engineers and geologists and will no doubt help to guide the profession for many years.

### Site models

Before or during the desk study phase of a civil engineering ground investigation (see Figures 4.1–4.3), engineering geologists, by virtue of their training and experience, should be able to visualize an initial simple basic model of any part of the Earth's surface or near-subsurface (a terrain model). We consider that geology taught to degree level is an essential part of the training of an engineering geologist. A geologist is trained to visualize the third (and fourth) dimension of any area. This visualization improves with experience. The details and accuracy of the model will depend on the location and the individual geologist.

Such models, especially in a new area, help us to understand the three-dimensional geology, near-surface variations and to identify the relevant geomorphological systems and processes and main environmental controls (see Parts 2, 3 and 4). Other disciplines may well be needed to help evaluate the situation – for example, geomorphology, which is indispensable at many sites, surface and groundwater hydrology, ecology, and specialist seismic and volcanic studies. This is why a geoteam is needed. Each site will require consideration of the approach to model-making and the value to be obtained. Interaction with the project's geotechnical engineers is essential at all stages of production of the model.

As an example, we can consider the construction of very long mineral railways over a large area in the Pilbara, Western Australia, for which little detailed geological mapping existed. In a series of investigations over a number of years, initial conceptual models were first developed to represent the nature of the different terrain units and engineering geological formations (Baynes *et al.* 2005). Observational models were then developed to present the observed and interpreted distribution of reference conditions in a variety of ways (maps, two- and three-dimensional models, sections). These were followed by evolutionary models, which illustrated the way in which the terrain units, engineering geological formations or the reference conditions had developed over geological time, using a series of sketch maps, sections and block models. The models therefore progressed from two to three to four dimensions.

Geomodels created during the desk study of a ground investigation will allow a better understanding of the site and more efficient planning of the investigation, such as the placing of boreholes, sampling and testing. The models should provoke both direct and lateral thinking and therefore lead to a more balanced and cost-effective investigation. The models are continually reviewed and improved as the investigation evolves. It is essential to continually check, refine or modify the initial model (see Part 4) so that it becomes increasingly site-specific. This must include continuous interaction of the geologists with the site and design engineers. Interaction is

also needed to make judgements on the management of the geo-risks associated with the project.

## Total geology

The total geological concept is required during the ground investigation stages following the early desk study to both develop thoroughly the initial geomodels and to present a comprehensive picture of the ground conditions (Fookes *et al.*, 2000). Unfortunately, this concept is not always foremost in an engineer's mind as a result of more pressing concerns, such as costs, time, quality and the availability of staff, equipment and materials. Three-dimensional hand or computer drawings must attempt to incorporate all the individual surface and subsurface components of the site, including the tectonic and structural geology, stratigraphy, geomaterials, ground and surface waters, the local climate and the geomorphological conditions, together with significant surface details resulting from human activities. For a small site, a model may only need a thumbnail sketch; for a larger site, a series of detailed models (of different geological aspects) are probably needed, particularly where future changes to the local environment and process systems, or a particular risk(s), are anticipated. The model must be 'engineer-friendly' – that is, easy to understand by those with little geological knowledge.

Other forms of model-making exist and these are being continually developed and becoming increasingly important, particularly through the use of computer-aided studies to evaluate large amounts of data (Culshaw, 2005; Allen *et al.*, 2014). These are not discussed here, but are now part of the geologist's toolkit.

### GEOLOGICAL TIME: STRATIGRAPHY

The age of Earth is of the order of 4600 million years (4600 Ma or 4.6 Ga) and Table 1 details the subdivision of this time to the present. The first fossils of simple cells date from about 3600 Ma and the oldest rocks found on the Earth's surface date from about 4400 Ma. Geologists divide the time from the birth of Earth to the present into long subdivisions or Eras and then into shorter Periods. Numerous other shorter time divisions

for specialized use are not shown in Table 1, which gives what we believe are the time names useful to engineers. Good indications of age can be given by the correlation of fossils in any sedimentary strata (e.g. the remains of shallow marine organisms present in a bed of limestone originally deposited on a shallow sea floor) or by radioactive dating. There are a variety of other ways to approximate the age of rocks by their type and history, such as dendrochronology or varve analysis and relative dating by the law of superimposition (Bibliography, Group A books).

### SURFACE MATERIALS

The current landforms, – that is, the subdivisions comprising the landscape in any region – are the result of the geological and geomorphological history. They may consist of several different types of rocks and soils, many of which will be familiar to geotechnical engineers. There are several good textbooks of engineering geology that describe such materials (Bibliography, Group B books) and can be read in conjunction with this book.

## Soils

Engineers generally consider any non-lithified (i.e. not rock-like) materials overlying solid rock (the bedrock) to be an 'engineering soil'. These are the 'overburden' of engineers or 'regolith' of geologists. The overburden may consist of saprolite (in situ weathered rocks and residual soils), described in Part 2, and/or a variety of soils transported by gravity, wind, water or glaciers before deposition (Table 2).

A small part of Earth's surface is bedrock that is more or less unweathered, or only a little weathered, as in some hot deserts or glaciated terrains. However, most of the world's bedrock is weathered and is covered by transported soils or soils developed in situ (e.g. tropical residual soils, temperate residual soils or duricrusts in arid lands; see Parts 2 and 3).

There are also special superficial ('surficial' in North America) coverings, which are described in more detail in Part 2. For example, the land surface of polar regions may be covered by ice, forest floors may be covered with decaying leaf

Table 1 Basic stratigraphic column showing the main divisions of geological time and the relative ages of major events (dates from the International Chronostratigraphic Chart, 2014).

| Eon | Era | Period | Age (Ma)[1] | Duration (Ma) | Orogenic[2] phases | Major biological, climatic and plate tectonic events |
|---|---|---|---|---|---|---|
| Phanerozoic (Evident life) | Cenozoic (Recent life) | Quaternary (Pleistocene and Holocene[3]) | 2.58 | Ongoing | Himalayan (ongoing) | Anthropogenic contribution to climate change and mass extinction — Major glaciations in the Northern Hemisphere |
| | | Tertiary[4] | 66 | 63 | Pyrenean Alpine-Laramide | First hominids — Age of mammals, birds and flowering plants — Extinction of dinosaurs considered to be caused by major meteorite impact and volcanism |
| | Mesozoic (Middle life) | Cretaceous | 145 | 79 | | Indian and southern oceans open as Gondwanaland breaks apart — First birds, modern bony fishes, rudist bivalves and flowering plants |
| | | Jurassic | 201 | 56 | Nevadan | Opening of North Atlantic Ocean — First mammals |
| | | Triassic | 252 | 51 | | First dinosaurs — Break-up of Pangaea into Laurasia (north) and Gondwanaland (south) |
| | Palaeozoic (Ancient life) | Permian | 298 | 46 | Uralian Hercynian-Appalachian (Variscan) | Mass extinction of rugose corals, trilobites and many other species — Glaciation in the southern hemisphere — Pangaea supercontinent formed — First reptiles — Last graptolites |
| | | Carboniferous (Pennsylvanian (Upper Carb), Mississippian (Lower Carb) in North America) | 358 | 60 | | |
| | | Devonian | 419 | 61 | Bretonian-Acadian Caledonian | First insects and amphibians — First land-living animals and plants |
| | | Silurian | 443 | 24 | | First fish with jaws |
| | | Ordovician | 485 | 42 | | Worldwide glaciation and mass extinction of marine life — First vertebrates (jawless fish) — First graptolites |
| | | Cambrian | 541 | 56 | Cadomian | First skeletal organisms — First soft-bodied animals, forming tracks and trails (about 900 Ma) — Increasing atmospheric $O_2$ (about 1.7 Ga) |
| Proterozoic | | Precambrian (all rocks older than Palaeozoic) | 2500 | c. 4000 | Penokean (Huronian) | |
| Archaean | | | 4000 | | Algoman (Kenoran) | Plate tectonic motions commence (about 3 Ga) — Earliest bacteria (about 3.5 Ga) — Major cratering on the Moon (about 4.2 Ga) — Oldest rocks on Earth's surface (about 4.4 Ga) |
| Hadean (Priscoan) | | | 4600 | | | Formation of Earth |

[1] Not to scale.

[2] Major plate tectonic phases.

[3] Also called Recent.

[4] Now commonly divided into Neogene (Upper) and Palaeogene (Lower).

litter and peaty organic soils may develop in wet areas. Surface clayey soils typically start as under-consolidated clays (see Appendix). The sea floor is not considered in this book.

## Rocks

Rocks of similar types occur in suites or associations. The long-term differential weathering and erosion of stronger and weaker rocks are reflected in the various landscapes that have developed around the world (see Parts 1, 2 and 3).

Rocks are loosely defined in geology as all forms of deposit that are older than the Quaternary Period and, using this definition, some forms of 'soil' may be called rocks. Many other firm or loose definitions of 'rock' and 'soil' exist for engineering situations, including the strength and difficulties of excavation. Engineers may call ancient clay sediments (e.g. over-consolidated London Clay, around 40 Ma old) an 'engineering soil' (see Appendix). This subject is fraught with potential contractual misunderstandings and unambiguous definitions are needed in any contract documents (Bibliography, Groups A and B books).

There are three main rock types.

- *Igneous rocks.* Intrusive igneous rocks solidified slowly from hot magma (molten rock generated by heating within the Earth's crust or upper mantle) before it reached the surface, forming large crystals, e.g. granites and dolerites. Extrusive rocks such as basalts cooled quickly from hot surface lava flows and have small crystals, whereas pyroclastic rocks formed from volcanic ejections through a volcanic vent, e.g. ash, cinder and larger debris (tephra), making tuffs, agglomerates and volcanic breccia. Intrusive and extrusive rocks tend to be strong and their behaviour may be governed by jointing systems. Pyroclastic rocks tend to be variable in their engineering performance, from weak and friable to strong and tough.

- *Metamorphic rocks.* These are rocks that have been altered by the effects of high confining pressures and/or high temperatures within the Earth's crust. Regional metamorphism is a result of both high temperatures and high pressures and is typically associated with mountain chains along plate collision margins (e.g. the Himalayas or Andes). Thermal or contact metamorphism results from the high temperatures around igneous intrusions, which bake the original rock. For example, mudrocks may be altered to shales, slates, phyllites or mixed schists depending on the temperature and pressure conditions. Metamorphic rocks can be broadly divided into three main groups: foliated (or banded), consisting of rocks in which the texture is layered, e.g. gneiss; those where the minerals have a preferred orientation, such as schists and slates; and those that are non-foliated – these rocks tend to have high isotropic strengths and low permeability, e.g. hornfels and granulite. Foliated rocks tend to be weak parallel to the planes of foliation.

- *Sedimentary rocks.* These are rocks formed from material derived from pre-existing rocks (i.e. sediment) and those of organic or chemical origin. They form a large part of the Earth's surface rocks. A distinctive feature of most sedimentary rocks is their stratification or bedding, which tends to control their behaviour. Clastic rocks are composed of particles or fragments that have been deposited from material derived from the weathering and erosion of pre-existing rocks. This is then lithified by compaction and cementation (diagenesis) at low pressures and temperatures to form mudstones, shales, siltstones, sandstones, gritstones, breccias and conglomerates. Chemical sedimentary rocks are typically formed from the precipitation of dissolved minerals, e.g. rock salt, gypsum and some limestones. Organic sedimentary rocks, such as shelly limestones, chalk and coal, are formed from the hard parts of animals and plants.

The mineral composition, fabric and porosity of rocks determine their mechanical strength and resistance to weathering. For example, shale is mechanically weak, but is resistant to chemical weathering, whereas limestones are often strong and resistant to mechanical weathering, but are readily soluble under slightly acidic conditions (chemical weathering), resulting in the formation of karst (cavernous) landscapes (see Parts 2 and 3; Bibliography, Group B books).

In general, the older rocks within a particular area tend to be stronger and have a more complex structure than younger rocks. Their geological history is also important and rocks of the same age in different locations (commonly in different tectonic settings) may have very different geological and engineering characteristics (see Parts 2 and 3; Bibliography, Group B books).

## THE QUATERNARY PERIOD: CLIMATE CHANGE AND THE GEOMODEL

The climate has varied throughout the history of our planet, over both geological and historical timescales. Examples of the most important worldwide variations are the four 'icehouse' phases, dominated by repeating glacial episodes, and the four 'greenhouse' phases dominated by repeating interglacial episodes, which occurred during the Phanerozoic Eon (Table 1). The slowly changing global distribution of land masses caused by plate tectonic movements has brought about either predominantly icehouse or greenhouse conditions. Each phase is believed to last about 40–95 Ma and greenhouse phases have accounted for about 60% of Phanerozoic time. There have also been special times within the Earth's history, such as part of the Cretaceous Period, when the Earth was in a long greenhouse cycle with no ice sheets over the poles, higher global temperatures and high levels of carbon dioxide in the atmosphere.

The last icehouse phase began in the middle Tertiary, building up to its greatest extent in the Quaternary. It was characterized by frequently occurring repeated climatic changes. Such repeated climatic changes with alternating interglacial and glacial episodes have been significant in the creation of today's landscapes, especially in the current regions of temperate climate. The ice advances and retreats within an icehouse phase are considered to be mainly a result of a combination of variations in the Earth's orbit around the Sun (the Milankovitch cycles) and changes in solar radiation.

In addition to the repeating ice advances and retreats of the Quaternary, smaller changes that have occurred

Table 2 Main types of transported soil.

| Soil type | Formation | Nature of deposit | Lithified [1] equivalent |
|---|---|---|---|
| Taluvium (coarse) | Transport down-slope mainly by gravity mass movement, e.g. talus and rock avalanche deposits (includes mountain soils) | Generally loose to poorly compacted, unsorted, unstratified weathered rock debris comprising angular gravel to very large boulder-sized material with various amounts of finer particles; typically deposited on steep slopes (25–35°) below cliffs or fault scarps; composed predominantly of strong rocks | Not commonly lithified, but fine to very coarse breccia, may be indurated with a clay matrix or cemented by calcium carbonate or iron oxides |
| Colluvium (fine) | Transport down-slope by combinations of gravity (creep), slope-wash and freeze–thaw action, e.g. debris and earth slides, sheet erosion (includes mountain soils) | Generally moderately compact, unsorted, unstratified, weathered rock debris dominated by clay-, silt- and sand-sized material, rarely with some angular gravel to boulders; typically deposited on moderate slopes (15–25°) where the underlying strata up-slope include a high proportion of weak mudrocks | Not commonly lithified, but matrix-supported ancient deposits may be indurated as mudstone or as a gap-graded fine to coarse sedimentary rock |
| Debris/earth/mud flow deposits | Rapid transport and deposition as a slurry by either overland or channelized flow (includes cold lahars, peat bog-bursts) | Unsorted, unstratified clay to boulders; often formed when saturated debris slides disintegrate; channelized flows may transport very large subangular or subrounded boulders; typically deposited on gentle slopes (5–15°) | |
| Solifluction deposits [2] | Slow down-slope movement of waterlogged soil material (includes mountain soils) | Variable; a type of colluvial soil, characteristic of gentle slopes in cold regions (gelifluction), but can occur on steeper slopes elsewhere | |
| Glacial | Transport and deposition by ice | Tills [3] of various types forming moraines, usually highly variable lithology; some tills are heavily over-consolidated | Tillite |
| Glaciofluvial | Transport and deposition by meltwater | Outwash materials, becoming finer away from the meltwater source; fine material usually laminated and varved (seasonal glacial lake deposits) | Claystone, mudstone, siltstone, sandstone, conglomerate and mixed sedimentary rocks |
| Alluvium | Transport and deposition by rivers | Fine clay to coarse gravels; coarse particles usually rounded; soils commonly sorted and often show pronounced stratification | |
| Aeolian | Transport and deposition by wind | Usually silts (i.e. loess) and fine to medium sands (e.g. sheet sands, sand dunes) with uniform grading; may be extensive | Well-sorted [4] siltstone or sandstone |
| Volcanic | Ash and rock fragments (i.e. tephra) and pumice, all deposited during eruptions | Silt to cobbles and small boulders; highly angular to subrounded, often vesicular; weathering of basaltic materials commonly produces highly plastic clays | Tuff, agglomerate, volcanic breccia |

[1] Lithified means 'turned to stone'.

[2] There are a number of regional names in Britain for solifluction deposits, e.g. head in southern England and coombe rock in the chalkland of south-east England.

[3] Older British geological maps may refer to till as boulder clay (this name is still commonly used by British engineers).

[4] Mainly single-sized material.

during historical times have influenced geomorphological processes and the Earth's surface systems on a regional scale. A geologist may need to incorporate some of this knowledge into geomodels (see Parts 1, 2 and 3; Bibliography, Group A books).

## Climate change

The legacy of climatic instability during the Quaternary Period, with its significant consequences for our current landforms, is a major geo-influence on today's engineering projects (see Parts 2 and 3, especially Figures 3.1, 3.2, 3.5 and 5.9).

Key features for engineering geologists to consider when making the initial geomodel include the following points.

- There have been marked global temperature fluctuations, from temperatures similar to those of the present day during interglacial periods to lower temperatures during the major glacial periods that were sufficiently cold to treble the volume of today's land ice. There have been at least 17 major glacial–interglacial cycles in the last 1.6 Ma.

- Immediately beyond the ice limits, permafrost and periglacial conditions (including tundra and taiga landscapes) have a profound effect on slope instability and cause near-surface freeze–thaw changes to the ground (see Figures 3.2, 3.5 and 5.9).

- Marked fluctuations in global sea level have included falls of more than 100 m during glacial periods, exposing parts of the continental shelves. Sea levels rose again during the warmer interglacial periods and flooded coastal regions to levels even higher than today. Important consequences of these regular rises and falls in sea level include 'buried' valleys cut below the level of modern rivers and subsequently infilled with sediments, complex river terrace sequences along valleys, onshore relict (former) sea cliffs, raised beaches, submerged forests, dead coral reefs and submarine canyons.

- Till sequences (formerly known as 'boulder clay') now cover much of temperate lands and nearby seabed surfaces. These were deposited from valley and continental glaciers that have now retreated to high latitudes and highland areas (see Figure 3.1).

- Glaciofluvial (or fluvioglacial) debris were deposited by rivers issuing from the margins of glaciers in vast volumes and often cover the tills laid down by glaciers as they advanced or retreated (see Figures 3.1 and 3.2).

- Numerous lakes of all sizes have been left in and near glaciated regions, many subsequently filled with seasonal deposits of laminated clay and silt or very unstable organic soils and peats (see Figures 3.1 and 3.2).

- Loess (predominantly wind-blown silt and fine sand) was exposed in great volumes as the glaciers retreated and was then carried away by the wind to cover vast areas of the Northern Hemisphere (see Figure 3.5).

- The rapid retreat and decay of the ice sheets over a few hundreds to a few thousands of years at the end of the interglacial periods replaced tundra landscapes with forests in the mid-latitudes. This is happening again today.

- In low latitudes, the growth and contraction of the high-latitude ice sheets corresponded approximately with periods of greater moisture (pluvials) and greater dryness (interpluvials). For example, during the Last Glacial Maximum (about 30,000–20,000 BP), the Ethiopian Highlands had a cold, dry climate with tundra at altitudes above 3000 m and glaciers in the higher mountains. The world's great sand seas developed during the dry periods and advanced into today's desert margins. Changes in monsoon patterns (partly associated with the rise of the Himalayas) also relate to changes in climate regimes.

- During the current interglacial period, starting around 15,000 BP, significant smaller climatic changes have been superimposed on the overall glacial–interglacial cycle.

For example, around 7000 BP sea levels were a metre or two higher than today and the Sahara experienced a humid period with extensive vegetation. The climate has continued to change over the last millennium – for example, the Medieval Warm Period (around 1100–1300 AD) was followed by the Little Ice Age (around 1550–1850 AD), probably due to variations in sunspot activity. Anthropogenic emissions of carbon dioxide and other greenhouse gases may be significant in today's climate changes.

## ENGINEERING AND THE LANDSCAPE

The impact of engineering on the landscape can also have significant effects on surface processes. Potential impacts need to be evaluated in relation to the site-specific conditions in the site geomodel. Although some impacts may appear to be localized, they can result in indirect consequences affecting the operation of surface processes throughout the whole surface system (Fookes *et al.*, 2007). The following points are examples of the impact of human activities on the landscape.

- Changes in erosion rates, such as the accelerated erosion seen in recently deforested upland areas and hill-slopes, notably during the 1930s 'dust bowl' of the south-western USA.

- Reductions in slope stability caused by artificial recharge of the local water-table – for example, by leakage from septic tanks or water supply pipes. Shallow landslides may be triggered by forestry logging activities, e.g. in British Columbia, Canada.

- Cut-slope failures may be caused by mining activities and excavations to create roads and level plots, e.g. a 94-fold increase in landslides on Vancouver Island, Canada has been associated with the construction of access roads across the forested mountain.

- An increase in, and acceleration of, run-off within urban areas may be caused by the construction of impermeable surfaces (e.g. concrete, tarmacadam and housing) within a catchment, e.g. at Stevenage, UK, the mean annual flood volume increased by 2.5 times after the construction of this new town.

- Numerous examples exist of changes in river discharge and flood behaviour, including sediment transport along rivers and changes in the delivery of sediments from river channels to flood plains and from rivers to the open coast.

## Geotechnical problems with engineering soils

The main engineering soils are detailed in the Appendix. This is in the form of a large table giving the typical geological and engineering characteristics of common soil types and the associated practical problems that may be encountered during ground investigations, construction and operation.

| Climate region | Morphoclimatic zone (Fig.1-2) | | Mean annual temperature (°C) | Mean annual precipitation (mm) | Relative importance of various geomorphological processes (see also Fig. 2.2) | Potential geohazards |
|---|---|---|---|---|---|---|
| Polar and Tundra (Frigid zone) | | Glacial (Fig.3-1) | -30 to -20 | 25 to 250 | Ice caps: high mechanical weathering rates, especially frost action; low chemical weathering rates; glacial action at a maximum; wind action significant; fluvial action confined to seasonal melt | Melting ice contributes to rising global sea level; outbursts of water and sediment may occur from beneath glaciers (e.g. jökulhlaups) |
| | | Periglacial (Figs.3-2 and 5-9) | -14 to -1 | 150 to 600 | Tundra: very active mechanical weathering with frost action at a maximum; low to moderate chemical weathering rates; high rates of wind action locally; seasonally active fluvial processes | Very active mass movements related to seasonal thawing of the active layer above continuous permafrost |
| | | Subarctic (Fig.3-2) | -7 to 2 | 150 to 600 | Taiga: severe winters with ice and snow storms; high mechanical weathering rates; low to moderate chemical weathering rates; permafrost developed under Pleistocene glacial and periglacial conditions and is not forming at present | Subsidence caused by thermokarst development in areas where discontinuous ice-rich permafrost, formed under colder Pleistocene climates, is currently degrading |
| Temperate and Mediterranean | | Humid mid-latitude maritime (Figs.3-4 and 3-5) | 5 to 20 | 400 to 1800 | Relict polar and tundra terrain: moderate chemical weathering rates, increasing to high at lower latitudes; moderate mechanical weathering rates with frost action important at higher latitudes; moderate rates of fluvial action; wind action generally confined to coastal areas | Moderate to high mass movement activity, particularly landslides on steep de-vegetated slopes; coastal erosion; sea, river and groundwater flooding |
| | | Seasonally wet mid-latitude continental (Figs.3-3 and 3-5) | 0 to 10 | 350 to 1000 | Relict polar and tundra terrain: rates of processes influenced by low winter temperatures, spring snowmelt and warm summers with severe thunderstorms: seasonally active mechanical weathering, especially frost action; low chemical weathering rates | Moderate mass movement activity, particularly landslides on de-vegetated slopes; seasonal river erosion and floods |
| | | Dry mid-latitude continental (Fig.3-3) | 0 to 10 | 100 to 350 | Semi-arid steppes: low to moderate chemical weathering rates; seasonally active mechanical weathering, especially frost action; rainstorms and active fluvial processes in summer; moderate rates of wind action locally | Episodic gullying, wind deflation and occasional sheet wash on ploughed soils and de-vegetated slopes |
| | | Mediterranean and hot semi-arid regions (Fig.3-4) | 10 to 25 | 200 to 600 | Alternating wet-dry conditions in areas bordering mountains, seas and deserts influence rates of all processes: moderate to low chemical weathering; active mechanical weathering at drier and cooler margins; episodes of high fluvial corrasion, dissolved and suspended loads; moderate to high wind action | Sporadic local landslides, sheet erosion, gullying and piping on de-vegetated or degraded slopes; downstream floods with high spatial and temporal variability |
| Tropical (Torrid zone) | | Hot deserts (Fig.3-6) | 10 to 25 | 0 to 200 | Arid lands: high mechanical weathering rates, especially salt weathering; minimal chemical weathering rates; generally very low but sporadically high rates of fluvial action; wind action at a maximum | Minimal rates of mass movement; dust storms; occasional flash floods |
| | | Wet-dry tropical (Fig.3-7) | 20 to 30 | 600 to 1500 | Savannah (grassland to scrub and woodland): active chemical weathering during wet season; low to moderate mechanical weathering rates; high rates of fluvial action during wet season, with overland and channel flow; generally minimal rates of wind action, but locally moderate during dry season | Moderate rates of mass wastage during the wet season, including landslides, debris flows, sheet wash and gullying; locally high seasonal rates of river erosion and avulsion |
| | | Humid tropical (Fig.3-8) | 20 to 30 | >1500 | Wet tropical rain forest climax vegetation: high chemical weathering rates due to high temperatures and high rainfall; very low mechanical weathering rates; moderate to low rates of fluvial corrasion but locally high rates of dissolved and suspended load transport | Active episodic mass movements, particularly debris slides on steep slopes; deforestation contributes to global climate instability |
| Azonal | | Mountains (Fig.3-9) | Highly variable | Highly variable | Rates of all processes vary considerably with altitude and location; mechanical and glacial action significant at high elevations; mountains and high plateaux influence global airflow patterns and affect the climates of adjacent regions | Rockfalls, avalanches, landslides, debris flows, flash floods (plus earthquakes and volcanism in tectonically active regions) |

*Figure 1.1 Characteristics of the major global morphoclimatic zones.*

# *Part 1. Underlying factors: climate and geology*

## *Characteristics of the major global morphoclimatic zones (Figure 1.1)*

Figure 1.1, which details the major global morphoclimatic zones and their relevance to geomorphological processes and potential geohazards, is used here as a general introduction. It should be considered in conjunction with Figure 1.2, a map of these morphoclimatic zones. Both the table and the map are approximate and should be treated as guides only. The table was developed from the map, which is based on Tricart and Caileux (1965) and Stoddart (1969). A variety of climate/soil classification maps that bring together different aspects of geological processes – for example, tectonism, sedimentation, weathering zones, basic geomorphology and soil processes – are discussed in Fookes (1997a).

Part 3 of this book is based on the three major climatic types that reflect global zonal variations in precipitation and temperature. These variations result from the unequal distribution of solar radiation, the global circulation of the atmosphere and ocean currents, and the relative positions of the continents and oceans. There is a fourth climatic zone, called azonal, which occurs in high mountains anywhere in the world: the higher up the mountain, the colder the conditions (see Figure 3.9).

- *Polar and tundra.* This climatic type is dominated by low solar radiation and cold temperatures. A polar climate is characterized by continuous low temperatures with perpetual ice and snow. The tundra environment is associated with permanently frozen ground (permafrost), but without permanent ice and snow, and lies between the polar regions and the northern (or southern) limits of tree growth (the taiga). In the Arctic, Antarctic and high mountains, precipitation is typically very low and low temperatures prevail throughout the year, seldom rising above 0°C (commonly called the frigid zone). Precipitation and temperatures are higher in the periglacial and subarctic environments; average temperatures here during the summer months are well above freezing. The subarctic is adjacent to, but outside, the Arctic Circle. Figures 3.1 and 3.2 illustrate glacial and periglacial environments.

- *Temperate and Mediterranean.* This climatic type is dominated by the westerly upper atmosphere jet stream, which controls the tracks of rotating low pressure regions. Variations in the atmospheric circulation can lead to changes in the patterns of drought, wet years and frequency of floods. The temperature relates largely to the amount of sunshine. A Mediterranean climate is characteristic of the western margins of continents in warm temperate zones, typically the mid-latitudes from 30° to 40°, with hot dry summers and cool moist winters. The term 'temperate' is from early geographical usage as the zone between the 'torrid' and 'frigid' zones; in modern usage it indicates climates with no great extremes. Figures 3.3 and 3.4 discuss the semi-arid temperate and temperate Mediterranean environments.

- *Tropical.* This climatic type is dominated by the atmospheric Hadley cells (inter-tropical convergence zones) consisting of areas of low pressure near the equator (the Equatorial Trough) towards which persistent winds (the trade winds) blow. Areas of subsiding dry air give rise to major hot deserts such as the Sahara. Intense rainfall, floods and/or drought events are associated with tropical storms, such as Atlantic hurricanes and Pacific typhoons, and with the monsoons. The El Niño Southern Oscillation causes abnormal atmospheric and environmental conditions, primarily in equatorial regions. It consists of two components: (1) El Niño (Christ child) is associated with strong fluctuations in ocean currents and surface temperatures within the Pacific Basin and (2) La Niña (little girl) is associated with abnormal, cold ocean surface temperatures in the equatorial Pacific Ocean. A tropical climate does not have a precise definition, but there is no cool season and no month has a mean temperature <20°C (the 'torrid' zone of earlier usage). Typical continental climates have pronounced wet and dry seasons. Figures 3.6 to 3.8 detail hot desert, savanna and hot, wet tropical environments, respectively.

Glacial environments are distinguished by the ice cover that dominates their processes of erosion and deposition. Ice caps and glaciers today extend over high mountain and polar regions, but large areas that now have temperate climates bear relict glacial landforms that were created during the Ice Ages in the cold stages of the Pleistocene.

Peripheral to the glaciers, the periglacial zone is cold enough to have permanently frozen ground (permafrost}, but snow melts away every summer, exposing the sparse plant cover of tundra or taiga. Conservation of the permafrost is essential to keep engineered structures stable on soils that derive their strength from the ground ice.

Temperate zones have gently rolling terrains defined by fluvial processes, where rivers cut sinuous, V-shaped valleys that evolve into wider and lower profiles. The typical result is a landscape with widespread small-scale modifications by human activity, and typically with large-scale features inherited from colder climates in the past.

The semi-arid and savanna zones are largely shaped by fluvial erosion and deposition, but processes are slower due to the lower rainfall. Long-term weathering and soil erosion leave residual masses of almost unweathered rock that form small, isolated hills, and gully erosion is widespread on steeper slopes with little or no plant cover.

The dry conditions in the hot deserts allow wind to be a major mechanism behind sediment transport, so that sand dunes and dust bowls are characteristic landforms. However, water erosion, that follows infrequent rainstorm events and flash-floods, is so powerful that dry valleys, wadis and fluvial sediments are features of all deserts.

The hot, wet, tropical zones are distinguished by their cover of dense rain forest, which invariably requires complete clearance prior to any engineering works. This exposes thick, red, clay-rich soils that are commonly unstable when their plant cover is removed, and stable bedrock is only found beneath deep weathering profiles.

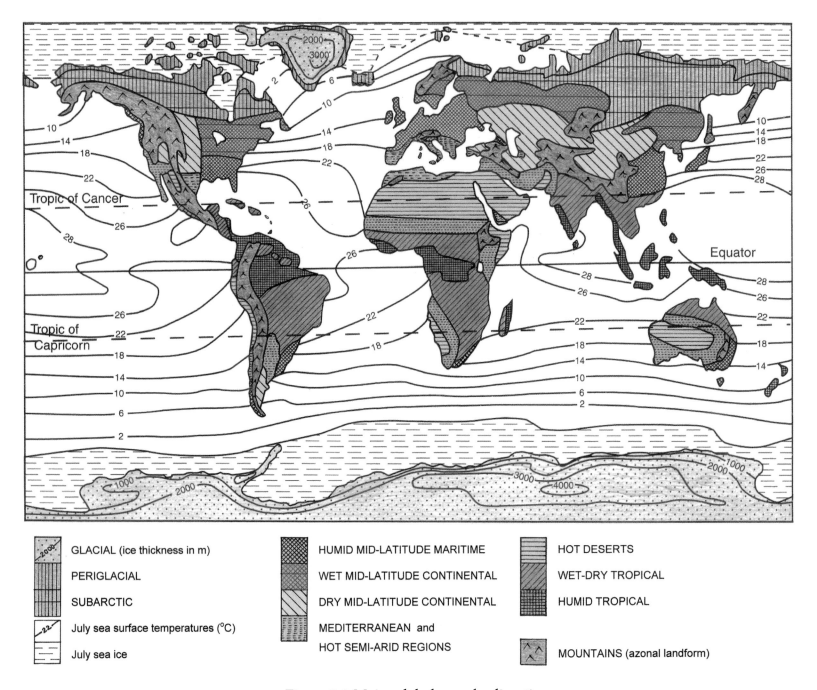

GLACIAL (ice thickness in m)

PERIGLACIAL

SUBARCTIC

July sea surface temperatures (°C)

July sea ice

HUMID MID-LATITUDE MARITIME

WET MID-LATITUDE CONTINENTAL

DRY MID-LATITUDE CONTINENTAL

MEDITERRANEAN and
HOT SEMI-ARID REGIONS

HOT DESERTS

WET-DRY TROPICAL

HUMID TROPICAL

MOUNTAINS (azonal landform)

*Figure 1.2 Major global morphoclimatic zones.*

## Major global morphoclimatic zones (Figure 1.2)

This map should be read in conjunction with Figure 1.1. It is a simplified map and the boundaries must be considered as approximate.

Climate is a major influence on the rate, scale and significance of the near-surface processes, weathering and erosion that dominate the construction of site models. In many regions, extreme climatic events (e.g. high intensity rainfall events) are responsible for initiating short-term changes in the landscape. Extreme events may also be responsible for a disproportionately large share of erosional changes. Catastrophism, the theory that associates past geological changes with catastrophic happenings, suggests that >90% of change takes place over <10% of time. This is valid in many environments and includes effects of considerable importance to communities and engineers, such as massive river erosion in major flood events and major landslides in young mountains, and large coastal changes induced by major storm waves.

- *Water* is the most critical factor in most aspects of landform change in all climates. The effects of water include the erosive power of rainfall and running water (even as rare events in deserts and in the narrow summer melt water window of glaciated regions), pore and cleft water pressure in soils and rock fractures, and the influence of groundwater through-flow on the weathering and leaching of soils and rocks (see Figure 2.1).

- *Frost action* can only occur in areas where the ground temperatures fall below freezing. Periglacial conditions occur outside the areas of total ice cover characterized by glacial landscapes and allow repeated freezing and thawing. This maximizes the frost-splitting of rocks, freeze–thaw ground movements (e.g. active layer detachment slides see Figure 5.9) and solifluction. Permafrost only develops when the ground remains below 0°C for at least two consecutive years. This only occurs in locations where the average air temperature is –2°C or colder (see Figures 3.1, 3.2 and 5.9).

- *Aridity* in hot drylands (e.g. hot deserts and semi-deserts) and cool drylands (e.g. the Gobi Desert) results in moderate to severe restrictions on the growth of vegetation. This, in turn, leads to reduced surface stability and a susceptibility to increased run-off when the occasional rainfall does occur, thereby increasing both the potential for erosion and flash flooding (see Figure 3.6).

- *Aggressiveness* is defined as a combination of high temperatures and low precipitation in hot drylands that results in net evaporating conditions. The downward leaching of salts within the ground is limited and even highly soluble salts such as sulphates and chlorides can remain in the soil profile. These produce a highly aggressive environment in which salt weathering is an important factor in rock disintegration and ground heave and in which concrete and roads can also be affected. Elsewhere, duricrusts may form from the predominant local surface material (e.g. calcrete in limestone terrains), especially in semi-arid conditions (see Figures 2.5 and 3.6).

- *Wind action* occurs worldwide, but is generally most effective in hot and cool drylands with a mean annual rainfall <200 mm. This results in the creation of mobile sand and dust hazards, such as dune formation and migration, and wind-scoured erosional features (see Figure 3.6).

- *Residual soils* are the product of the *in situ weathering* of the bedrock. The soil thickness and type are broadly associated with the climate of the region (temperature and precipitation), which together influence the intensity of weathering, especially in temperate and tropical environments. The depth of weathering reflects the relative balance between the rate of bedrock weathering and the rate of removal by soil erosion or landslides. Maps of the distribution of the different types of residual soils are helpful in a first engineering evaluation (Fookes 1997b; see also discussions in Part 2, especially Figure 2.2).

- *Tropical residual soils* usually exhibit a distinct engineering strength, bearing characteristics and other properties. Weathering may exceed 100 m in depth, especially in humid tropical environments, and will normally be highly irregular in both depth and profile, leading to difficult differential foundation conditions (see Parts 2 and 3, Figure 3.8).

13

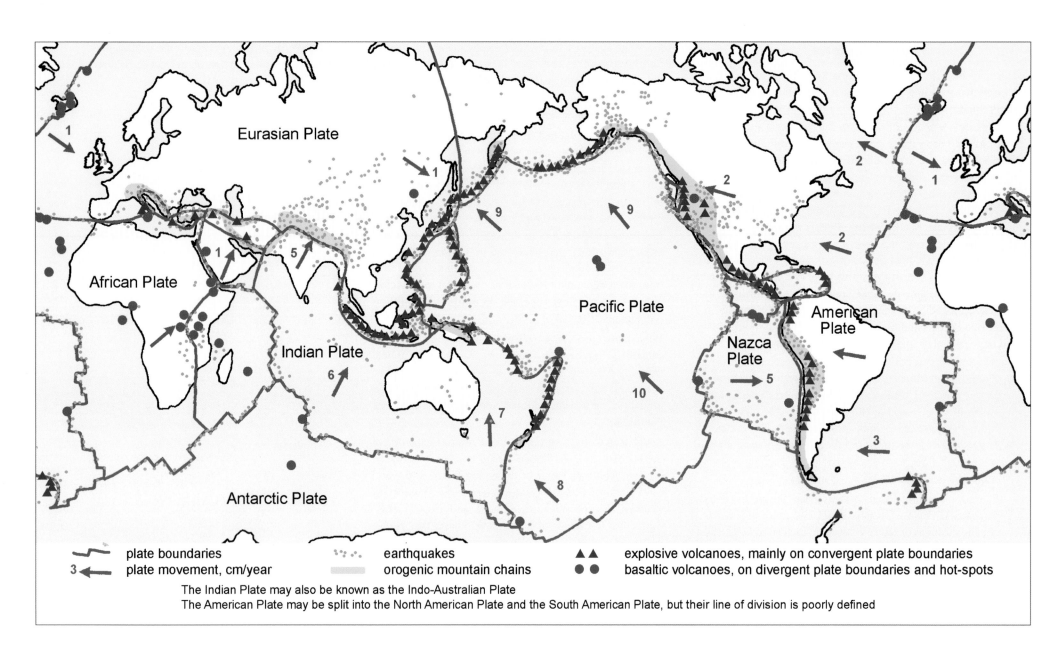

*Figure 1.3 Seismo-tectonic and volcanic activity related to current tectonic plate boundaries.*

## Major movements of crustal plates: volcanic and earthquake zones (Figure 1.3)

Landforms and their near-surface geology are ultimately the result of geological and geomorphological processes that derive their energy from plate tectonics and the climate. Plate tectonic movements – which include mountain-building, crustal warping, earthquakes, volcanism, folding, faulting and metamorphism – are a result of subsurface heat flow. The effects of this activity are important in many parts of the world, including the circum-Pacific volcanic/earthquake belt and areas of geologically young orogenic mountains, such as the Himalayas and the Andes, which are still growing (see Figure 3.9).

The theory of plate tectonics, formulated during the late 1960s, unifies the processes of continental drift, sea-floor spreading, mountain-building, seismic activity and volcanism into a single coherent model. It was incorporated into taught geology courses and textbooks in the 1970s.

Plates are formed in the lithosphere, which is the relatively brittle outer rock layer consisting of the crust and upper mantle of the Earth. Plates are huge broken slabs of the Earth's crust. The outer crust is up to about 100 km thick and consist of the thinner oceanic crust and the thicker continental crust made from various solid silicate rocks (silicates are the most abundant of the rock-forming minerals). The mantle, about 2800 km thick, consists mainly of hot iron silicates, which behave in a plastic manner. Convection currents circulate in the mantle as a result of the heat produced by the largely molten iron–nickel core at the centre of the Earth. These convection currents form cells in which the hotter material rises and the cooler material falls, with horizontal movement at the top of each cell immediately underneath the crust. The plates move very slowly, only a few centimetres per year, dragged by the flow of the underlying convective mantle. The direction of movement of the plates is indicated on Figure 1.3 by arrows; Table 1.3.1 summarizes the characteristics of the plates.

The plates are relatively stable. However, major disturbances occur at their boundaries where the plates collide, diverge or slide past one other, causing most of the active geological processes encountered on Earth. These include the formation of new igneous, metamorphic and sedimentary rocks, their deformation and mineralization, with severe folding, faulting and subsequent erosion.

In the unstable areas at the plate boundaries, additional energy is provided to alter the rate of geomorphological and geological processes. As an illustration, young mountains in an upward-moving phase are sites of landslides, erosion and the down-cutting of rivers, which typically become more active to keep pace with the uplift. Similarly, there are upwards or downwards changes in ground level relative to sea level. Examples of the effects of changes in sea level include the risk of coastal inundation, the rearrangement of coastal features, down-cutting of the long profile of rivers, or the burying of former deep channels. An understanding of plate tectonics is therefore important for geologists constructing site models. Fortunately, there are many good geological textbooks available (e.g. Waltham, 2009; see also Bibliography, Group A books).

The ocean crust is 5–10 km thick and forms ocean floors composed mainly of basic basaltic and doleritic igneous rocks. It is created at divergent boundaries by upwelling and destroyed at convergent boundaries by subduction under the thinner of the two colliding plates. The continental crust (20–80 km thick) is mainly made of acidic granitic and gneissic rocks, which are of a lower density than the rocks of the oceanic crust. The continental crust therefore floats higher on the mantle, forming all the continents, submerged continental shelves and adjacent islands. This crust is usually being eroded, but may be added to by the accretion of sediments and rocks scraped off any subducting oceanic plate.

There are three principal types of plate boundary (also see Table 1.3.1).

- *Divergent boundaries* are constructive boundaries where new ocean plates are formed and subsequently diverge as a result of the upwelling of lighter basaltic magma (very hot fluid rock) created by silicate liquids separating from heavier iron-rich solids in the partly melted mantle. This process produces numerous igneous dykes and submarine volcanoes and also creates mid-oceanic islands from the excess magma – for example, Iceland, which lies on the line of the Mid-Atlantic Ridge.

- *Convergent (collision) boundaries* are destructive boundaries where one plate is subducted below the other and undergoes melting. The most important type of collision boundary is one where a thinner oceanic plate is subducted beneath a thicker continental plate. This collision produces a great range of geological processes, collectively known as orogenesis. Where plates collide, the over-riding plate is crumpled and thickened to form a mountain chain. Probably the clearest example is where the eastwards moving Pacific ocean floor is subducted beneath the South American plate to form the geologically active Andes mountain chain.

- *Transform (conservative or shear) boundaries* are where two plates meet and slide past each other in a sideways movement, e.g. the San Andreas Fault. Major tear faults (see Figure 1.10) are formed and intermittent movement creates major earthquakes.

As the major processes in plate tectonics evolve and the patterns of the plates change over geological time, new oceans open up, continents collide and may be welded onto one another. Therefore, any one place on the Earth's surface is a product of the different geological environments that have occurred throughout geological time. Time is thus the fourth dimension that must be appreciated to understand the geology of any area and in the construction of site geomodels.

Ancient orogenic mountains (formed hundreds of millions of years ago) are the products of plate collisions that have now ceased. Their high lands are being constantly worn down and their slopes are becoming flatter and more stable, e.g. the Highlands of Scotland, the mountains of Wales and the moorlands of south-west England. Eventually, after a very long period of geological time, an essentially flat, stable landscape with a deep weathering profile is produced, unless

Table 1.3.1 Characteristic features of active seismo-tectonic and volcanic regions.

| Tectonic zone | Occurrence and general characteristics | Relative importance of seismic and volcanic processes | | Particular geohazards |
|---|---|---|---|---|
| | | *Seismic activity* | *Volcanic activity* | |
| Divergent plate boundaries | *Constructive margins*: tensional stress initially forms a continental rift valley (e.g. the main Ethiopian Rift), which develops in stages through sea-floor spreading (e.g. the Red Sea) to a mature ocean basin (e.g. the Atlantic Ocean); new oceanic crust is generated as the plates move apart at mid-ocean ridges (e.g. the Mid-Atlantic Ridge) | Seismically active continental rift zones (e.g. the East African Rift) experience damaging earthquakes on moderately to steeply dipping normal (extensional dip-slip) faults; seismic activity is generally low on the spreading sections of mid-ocean ridges, but much higher along their transform offsets | Basaltic volcanism derived from rising mantle magma. Basaltic lavas advance slowly and tephra-falls and gas releases are generally minor, so there is little threat to human life, although built structures may be destroyed; extensive late Cretaceous–early Tertiary continental flood basalts (e.g. the Deccan Plateau in India) are implicated in the Cretaceous–Tertiary mass extinctions | Alternating layers of strong basalt and weak scoria with palaeosols and weathered ash horizons cause cut-slope instability, excessive over-break in tunnelling and foundation problems; heavy rainfall may trigger landslides along faulted rift margins; under warm, wet conditions basic volcanic rocks weather to expansive smectite-rich vertisols (e.g. black cotton soils) that cause slope instability and foundation problems; the eruption of lava beneath glaciers causes destructive melt water floods (e.g. jökulhlaups in Iceland) and tectono-magmatic fissures on rift floors can disrupt transport routes |
| Convergent plate boundaries | *Destructive margins*: compressional stress moves two plates towards each other; thinner oceanic crust slides beneath thicker continental crust at a subduction zone and accretionary wedges are uplifted to form mountain ranges (e.g. the Andes); when two oceanic plates converge, one is usually subducted beneath the other, forming a deep ocean trench and island-arc volcanoes<br>*Collision zones*: when two relatively light continental plates converge, neither plate is subducted and the crust buckles to form mountains (e.g. the collision of India and Asia to form the Himalayas and the Tibetan Plateau); many mountain chains are the product of plate collisions that occurred between the Mesozoic and the present | Seismically and tectonically highly active, with very large, damaging earthquakes on shallow to moderately dipping thrust faults; the world's largest earthquakes occur in subduction zones such as that underlying the circum-Pacific orogenic belt; potential reactivation of the Cascadia Fault off the coast of British Columbia/north-west USA is currently causing concern; eruptions of mud, liquid and gas from vents (mud volcanoes) are often associated with subduction zones and orogenic belts | Explosive andesitic volcanism caused by accumulated gas pressure; dominated by pyroclastic flows, air-fall tephra and ash (e.g. the Pacific 'ring of fire'); lateral blasts, flank collapses, nuées ardentes, mudflows and lahars pose a great threat to life and built structures; sulphurous gas and dust emitted from a super-eruption in Sumatra 73,500 years ago is thought to have cooled worldwide temperatures by 3–5°C; volcanic activity is not normally associated with continent–continent collisions | Explosive eruptions create very variable ground conditions in which strong rhyolitic lavas and welded ignimbrites alternate with loose tephra and pyroclastic flow deposits and may obscure pre-existing weak strata, causing slope instability and foundation problems; major erosion and landslides occur in the young circum-Pacific and Alpine–Himalayan mountain belts; andesites may weather to unstable halloysite-rich andosols in warm wet–dry climates; clouds of fine erupted ash may disrupt air travel and large mud volcanoes may destroy housing and farmland (e.g. Lusi, Java in 2006); strong ground-shaking can trigger rock-falls, toppling failures, large rotational slides and liquefaction of susceptible fine soils. Highly destructive tsunamis are typically generated by sudden large displacements on offshore faults |
| Transform plate boundaries | *Conservative margins*: shear stress moves two plates laterally past each other; lithosphere is neither created nor destroyed; most transform boundaries offset the spreading sections of mid-ocean ridges, but some occur on land (e.g. the San Andreas Fault in the western USA and the North Anatolian Fault in Turkey) | Seismically active, with large damaging earthquakes on near-vertical transform (strike-slip) faults; the San Andreas Fault links two divergent boundaries, whereas the North Anatolian Fault links two convergent boundaries | Volcanic activity is not normally associated with transform plate boundaries | Strong ground-shaking can trigger rock-falls, rotational landslides and earth flows; the Lisbon earthquake and tsunami of 1755, generated by movement on the Azores–Gibraltar transform fault, caused devastation in southern Portugal and led to the birth of modern seismology |
| Areas remote from plate boundaries | Unforeseen release of stress within tectonic plates and volcanic eruptions above magma hot-spots | Occasional intra-plate earthquakes occur in otherwise stable continental regions (e.g. New Madrid, USA in 1811–1812; Guinea, west Africa in 1983; Gujarat, India in 2001) | Major volcanic activity occurs above hot-spots (e.g. Hawaii); Yellowstone caldera in the USA is thought to be a potential location for a very large volcanic eruption in the near future | Unexpected slope and foundation failures and collapse of built structures that have not been constructed to withstand earthquakes; sudden out-gassing of carbon dioxide from crater lakes can result in human and animal casualties (e.g. Lake Nyos on the Cameroon Volcanic Line in 1986) |

some other process, such as a continental ice sheet, removes the ancient weathering products. Very ancient areas in the middle of continents are called cratons (or shields), e.g. the Mid-African Craton and the Laurentian Shield of Canada. Younger land masses may have been welded onto their edges to create units of a continental scale.

The Himalayan mountain chain (Mount Everest is towards the left) that lies along the convergent boundary between the Indian and Eurasian plates.

Uplifted oceanic crust is formed of basaltic lavas and dykes that fill the gap created by the plate divergence; it lies above sea level on the volcanic pile of excess material that has formed Iceland.

Divergent plate boundaries are mainly on ocean floors, but the Mid-Atlantic Ridge lies across Iceland. This fissure in old basalt lavas is slowly widening and steam rises from a water-table pool in the geothermally warmed zone.

Conical composite volcanoes made of andesite lavas and tephra form an island chain along an active convergent boundary where one oceanic plate is being subducted beneath another.

Basaltic pillow lavas extruded underwater form oceanic crust exposed on land in a zone of major tectonic uplift (n.b. scale person).

Continental crust of granitic rocks forms this basement of high-grade gneiss, part of the Laurentian Shield.

Very old metamorphosed ironstones that were strongly folded within an orogenic belt.

[above] Between the Pacific and American plates, slow, transform fault creep on a branch of the San Andreas Fault deforms this street without causing any major earthquakes.

[left] Astride the main San Andreas Fault, this fence was offset by the sudden movement that caused the 1906 earthquake in San Francisco.

Three stages in the progressive erosion of mountain chains.

[top] Young orogenic mountains of the Karakoram form a range of high jagged peaks with thousands of metres of local relief.

[above] Older orogenic mountains in eastern Scotland have been eroded down to create a more subdued upland landscape.

[below] Very old orogenic rocks in Western Australia have been worn down to form a shield with low relief and at low altitude.

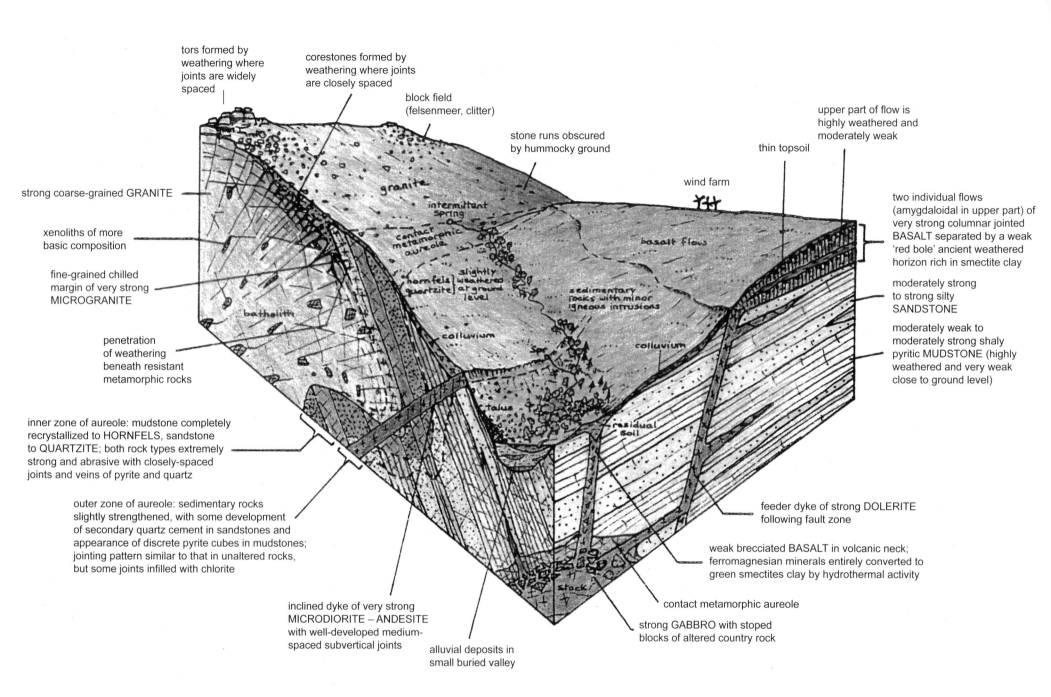

*Figure 1.4 Igneous rock associations.*

tors formed by weathering where joints are widely spaced

corestones formed by weathering where joints are closely spaced

block field (felsenmeer, clitter)

stone runs obscured by hummocky ground

upper part of flow is highly weathered and moderately weak

thin topsoil

wind farm

strong coarse-grained GRANITE

two individual flows (amygdaloidal in upper part) of very strong columnar jointed BASALT separated by a weak 'red bole' ancient weathered horizon rich in smectite clay

xenoliths of more basic composition

fine-grained chilled margin of very strong MICROGRANITE

moderately strong to strong silty SANDSTONE

moderately weak to moderately strong shaly pyritic MUDSTONE (highly weathered and very weak close to ground level)

penetration of weathering beneath resistant metamorphic rocks

inner zone of aureole: mudstone completely recrystallized to HORNFELS, sandstone to QUARTZITE; both rock types extremely strong and abrasive with closely-spaced joints and veins of pyrite and quartz

outer zone of aureole: sedimentary rocks slightly strengthened, with some development of secondary quartz cement in sandstones and appearance of discrete pyrite cubes in mudstones; jointing pattern similar to that in unaltered rocks, but some joints infilled with chlorite

feeder dyke of strong DOLERITE following fault zone

weak brecciated BASALT in volcanic neck; ferromagnesian minerals entirely converted to green smectites clay by hydrothermal activity

inclined dyke of very strong MICRODIORITE – ANDESITE with well-developed medium-spaced subvertical joints

alluvial deposits in small buried valley

contact metamorphic aureole

strong GABBRO with stoped blocks of altered country rock

## Igneous rocks (Figure 1.4)

Igneous rocks are formed when hot molten rock (magma) solidifies. Magma consists of hot solutions of several liquid rock phases, the most common of which is invariably a complex silicate. Hence igneous rocks are mainly composed of silicate minerals. Primary magma is generated by local heating and melting of rocks within the Earth's crust and upper mantle (see Figure 1.3). Depending on their composition, rocks melt at temperatures between about 500 and 1500°C. The temperature of melting also depends on the pressure, which is largely related to the depth at which melting occurs, and the water content. Differentiation into separate magmas and, ultimately, different rock types occurs when different minerals within the melt crystallize at different temperatures. When crystals form at high temperatures, the composition of the remaining liquid magma is changed and so a process of fractional crystallization occurs. This produces a series of rocks that are different from the original magma. Acidic (granitic) and basic (basaltic) or ultrabasic lavas, and many others between these extremes, can be formed. For further discussion, see Bibliography, Group A books, and Francis (1993).

When the magma cools, it solidifies into a crystal mosaic of minerals to form an igneous rock. The form and occurrence of igneous rocks depend on their structure in or on the ground. Acidic and basic lavas may cool over hours or days, whereas a granite batholith may take millions of years to crystallize. The chemical composition of igneous rocks is determined by the melt forming the original magma. The crystal size and degree of interlocking largely relate to the rate of cooling (the slower the cooling, the larger the crystals) and ultimately determine the strength of the rocks.

There are several different classification systems for igneous rocks, but this discussion is left to more specialized textbooks. Essentially, igneous rocks can be classified as either extrusive or intrusive. Intrusive rocks crystallize underground within the Earth's crust, whereas extrusive rocks are formed when the magma solidifies at the surface.

### EXTRUSIVE IGNEOUS ROCKS

Volcanoes produce extrusive rocks, but intrusive rocks also solidify inside and beneath them. When the underlying magma is rich in silica and viscous with a high gas pressure, volcanic eruptions are usually spasmodic, violent and explosive. Silica-poor magmas are very fluid and create quiet eruptions of more basic lavas. There is a range of igneous rocks between these two extremes. Pyroclastic rocks are extrusive rocks created when fine, fragmented material (tephra) is thrown into the air by explosive activity within the volcano. Much of this tephra is cooled in flight and lands as various types of pyroclastic rock, the names of which depend on the grain size and range from ash (fine-grained) to tuffs and agglomerates (coarse-grained). Hot tephra erupted under high temperature conditions may form pyroclastic flows, which can roll down the flanks of volcanoes and coalesce into welded tuffs (Francis, 1993).

### INTRUSIVE IGNEOUS ROCKS

Not all igneous rocks are produced from volcanic activity. Major intrusions include plutonic (deep-seated) batholiths, stocks and bosses, which are very large and are generally composed of granite and related rocks. For example, the Coast Range Batholith of Alaska and adjacent Canada is exposed over a length of 1000 km and is between 130 and 190 km in width. Batholiths are associated with orogenic belts along convergent plate boundaries. They are more or less stratified, but are not bottomless and may bake their host country rocks for some distance from the edge of the intrusion to form an aureole of contact metamorphic rocks (Figure 1.4; see also Figure 1.5).

Igneous intrusions are named by their geometric form, especially minor intrusions such as dykes and sills.

- *Dykes* are discordant intrusions – that is, they cut their host rock at a high angle and are therefore steeply dipping, often vertical. As a consequence, their surface outcrop is little affected by topography and they usually strike in a more or less straight line, e.g. the Cleveland Dyke in the north of England, which can be traced across the country for about 200 km. Large dykes may have irregular offshoots and all dykes may act as feeders for surface lava flows or subsurface sills. Most dykes are of a basaltic composition and they may be multiple or composite if they are formed by two or more injections of magma at different times. They can be up to tens of metres wide, but are usually less than 5 m wide. They may bake the host rock into which they are intruded (in zones a metre or so wide) and may have 'chilled' margins themselves. Unweathered dykes are typically strong rocks, but are often jointed, forming good foundations and construction rock in fill, armour and masonry.

- *Sills*, like dykes, are generally thin and parallel-sided intrusions and they usually occur over extensive areas. However, their thickness can be very large – up to several hundred metres. Unlike dykes, they are injected in a roughly horizontal direction, but, if subsequently folded, their attitude takes the form of the fold. Like dykes, they may bake the host rock into which they are intruded over a small distance and have 'cooled' margins themselves. Unweathered sills have engineering characteristics similar to dykes.

Much more could be written on the mineralogy, form and engineering characteristics of dykes, sills and their many related forms (e.g. laccoliths and phacoliths) and on the many types of volcano and volcanic rocks (e.g. shield volcanoes, cinder cones, pahoehoe lava), but such a specialist study is not appropriate for this book. For further reading, see Bibliography, Group A books, and Francis (1993).

[above] The Shiprock plug, in New Mexico, was a volcanic vent and now stands 500m tall due to faster erosion of the weaker surrounding rocks; a thin radial dyke from it now forms a wall of strong dolerite.

[below] A quarry in a granite batholith has faces cut by wire saws to extract stone that is free of joints.

Small intrusions of dolerite. [above] The Whin Sill in northern England intruded as a sheet into sedimentary rocks. [below] A swarm of dark dykes intruded into pink granite in a tectonic zone of tension; face is 15m high.

A Strombolian eruption from an active vent, and a dolerite plug that solidified within the vent of an ancient volcano.

[above left] A pyroclastic flow rolls down the flank of a volcano, and [centre] the deposits left by earlier flows.

[above right] The ruins of Pompeii that have been excavated from the flow deposits that buried the city in the AD79 eruption of Vesuvius (in the distance).

A dome of black andesite lava lies steaming in the crater of a large volcano ten years after its eruption, with a similar conical volcano beyond.

Columnar joints formed by cooling shrinkage in a basalt lava 5 m thick.

On Hawaii, a new lava becomes pahoehoe as its skin cools and wrinkles [above], and an old black pahoehoe flow lies on top of a red aa flow [below].

Variation in basaltic ground, in a road cut in Hawaii, with a solid, strong lava overlying a weaker lava with a rubbly aa top and an open lava tube inside it.

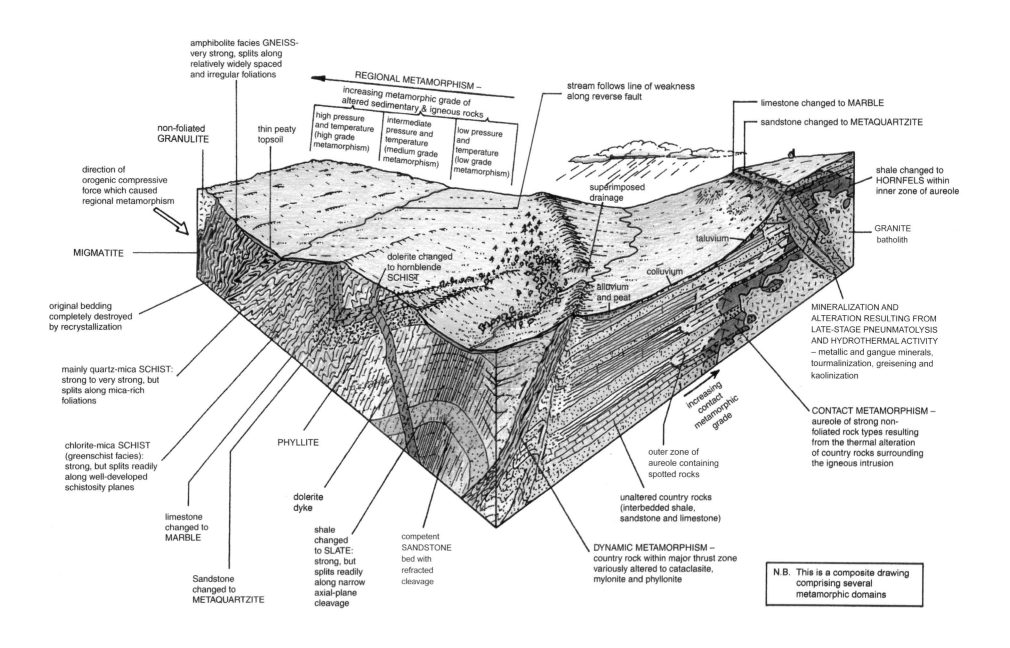

*Figure 1.5 Metamorphic rock associations in a wet temperate climate (after Fookes 1997a).*

24

## Metamorphic rocks (Figure 1.5)

Metamorphic rocks are derived from pre-existing rocks that have undergone mineralogical, textural and structural changes. These changes have been brought about within the physical and chemical environment in which the original rocks evolved by progressive transformation in the solid state. Changes in temperature and/or pressure are the primary agents causing metamorphic changes. Individual minerals are only stable over limited temperature and pressure ranges and, when these limits are exceeded, mineralogical adjustments are needed to establish new equilibrium conditions within the new environment. The term 'grade' refers to the range of temperatures under which metamorphism occurs.

Temperature increases to about 600°C and/or high pressures up to around 500 MPa (conditions typically occurring about 20 km below the ground surface) induce various changes, including the following.

- The rocks may recrystallize to form a strong mineral mosaic. The rock becomes stronger as a result of the strong interlocking of the individual grains.

- New minerals are grown at the expense of the original minerals as these become less stable under the new conditions of increased temperature and pressure. The most important changes are in the clay minerals. These minerals first transform to micas and then to feldspars, together with some mafic minerals (dark-coloured silicates), as the metamorphic grade rises. Micas are the most common minerals in metamorphic rocks and change to feldspars and mafic minerals only at the highest grade of metamorphism. Low-grade metamorphism usually produces chlorites and epidotes, both of which are distinguished by their green colour.

In high-pressure environments, structural changes and the reorientation of minerals occur in relation to the regional pressures. New minerals grow in the direction of least resistance, perpendicular to the maximum compressive pressure, to create a foliation or banding in the new rocks. Planes of weakness are created by the parallel alignment of minerals.

Fracture cleavage (shear cleavage) occurs where the cleavage planes are not controlled by mineral particles in parallel orientation. Refractured cleavage refers to the change in orientation of the cleavage planes as they pass through beds of different rock types. Slaty cleavage develops in fine-grained rocks as a result of intense deformation causing the partial recrystallization of platey minerals (mainly micas) parallel to the axial planes of folds – that is, perpendicular to compressive forces. Flow cleavage is a further development of slaty cleavage, finally leading to schistosity when no trace of the original bedding remains. Foliated rocks tend to split preferentially along the foliation planes when loaded or unloaded; the geometry of the failure planes is therefore an important aspect of rock stability in engineering.

- Non-foliated metamorphic rocks tend to be stronger. These rocks include: hornfels, which is formed by the thermal metamorphism of clay without high pressure; marble, a metamorphosed recrystallized limestone; and granulite, a high-grade metamorphic rock with little or no mica. Gneiss is a banded metamorphic rock that results from the separation of dark- and light-coloured minerals.

### TYPES OF METAMORPHISM

*Regional metamorphism* occurs over extensive areas in mountain chains formed by collision along plate boundaries. The rocks are subjected to both high temperatures and high pressures and regional metamorphism takes place when the confining pressure is typically in excess of 200 MPa. Lower pressures with high temperatures causes contact metamorphism. The maximum temperatures in regional metamorphism are about 800°C – most rocks melt above this temperature.

Slates are the product of the low-grade regional metamorphism of clay-rich sediments. As the metamorphic grade increases, these slates change to phyllites and then to mica schists, all of which are notable for their low shear strength parallel to the cleavage or schistosity (beware of instability in steeply dipping cuts in these rocks). Sandstones change to strong and hard quartzites, but quartz mica schists are formed if there is sufficient mica in the original sandstone. Quartzites may have strained molecular lattices and these rocks may be alkali silica aggressive in concrete (see Figures 5.6 and 5.7). Carbonate rocks simply recrystallize to marbles; if the original rock was dolomitic (a calcium/magnesium carbonates), then these marbles are usually yellowish or buff (see Figure 1.9). Basic rocks (e.g. basalt) are converted into strong greenstones or weaker greenschists; at higher metamorphic grades they first form strong amphibolites, then granulites and, finally, eclogites under very high temperatures and pressures.

*Thermal (contact) metamorphism* only involves high temperatures and typically occurs in metamorphic aureoles up to about 2 km wide surrounding large igneous intrusions. These igneous intrusions bake the original country (host) rock into which they were intruded. Aureoles may be sequentially zoned as the temperature decreases away from the intrusive body, producing strong rocks with few joints. Thermally metamorphosed limestones may form marbles, sandstones tend to have their quartz grains recrystallized to form quartzites and igneous rocks are usually only affected by very high grades of metamorphism and tend to produce gneissic rocks.

*Dynamic dislocation or cataclastic metamorphism* is, arguably, not true metamorphism because it gives rise to deformation rather than transformation to new rock types. It is only produced on relatively small scales, is usually highly localized and takes place in association with large faults or thrust movements. On larger scales, it is associated with big folds. Such processes produce crushing of the country rocks, seen as brecciation, granulation, mylonization, changes in pressure solutions, partial melting and slight recrystallization. In extreme cases the resulting crushed material may be fused to form a vitrified rock. Rock failures tend to occur preferentially on the planes of crushing.

For a fuller appreciation of this complex subject, see Bibliography, Group A books. For the engineering characteristics of metamorphic rocks, see Bibliography, Group B books.

High-grade metamorphic rock. A folded, foliated and banded gneiss with zones of paler migmatite, now forming the basement of an ancient shield.

Best-quality slate being quarried in North Wales, where the cleavage is nearly vertical and is crossed by at least three sets of fractures.

[above] Metamorphic schists strongly folded within an old orogenic belt.

[right] The giant heads of Mount Rushmore carved into a granite with faint banding that shows it has metamorphic origins.

[below] A stack of Welsh slates ready for use on a roof, after each was split along its natural cleavage to form the thin sheet, which was then guillotined to correct size.

Blocks of marble newly cut from a quarry face, where the mountain slopes of bare rock beyond appear similar to terrains on any other strong limestone.

The base of a dolerite sill (at shoulder height on the person), with underlying shales (down to waist height) thermally metamorphosed to dark hornfels, above pale, baked sandstone.

[above] Mineralised granite largely altered to clay by reaction with steam during low-grade metamorphism.

[left] Fumaroles within a geothermal basin, where the steam is also altering the rocks unseen within the ground.

A recent landslide in a mountain environment where metamorphic rocks have failed on the steeply inclined weakness formed by their schistosity.

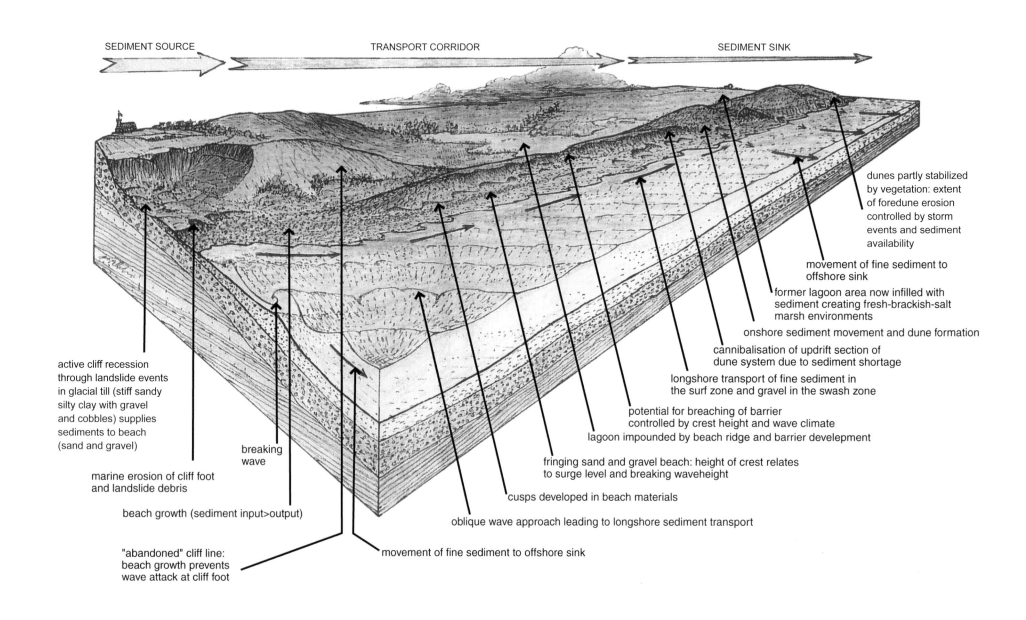

SEDIMENT SOURCE

TRANSPORT CORRIDOR

SEDIMENT SINK

dunes partly stabilized by vegetation: extent of foredune erosion controlled by storm events and sediment availability

movement of fine sediment to offshore sink

former lagoon area now infilled with sediment creating fresh-brackish-salt marsh environments

onshore sediment movement and dune formation

cannibalisation of updrift section of dune system due to sediment shortage

longshore transport of fine sediment in the surf zone and gravel in the swash zone

potential for breaching of barrier controlled by crest height and wave climate

lagoon impounded by beach ridge and barrier development

fringing sand and gravel beach: height of crest relates to surge level and breaking waveheight

cusps developed in beach materials

oblique wave approach leading to longshore sediment transport

movement of fine sediment to offshore sink

active cliff recession through landslide events in glacial till (stiff sandy silty clay with gravel and cobbles) supplies sediments to beach (sand and gravel)

breaking wave

marine erosion of cliff foot and landslide debris

beach growth (sediment input>output)

"abandoned" cliff line: beach growth prevents wave attack at cliff foot

*Figure 1.6 Sediments and sedimentary rocks: introduction and the coast.*

## Sediments and sedimentary rocks: introduction and the coast (Figure 1.6)

This figure and Figures 1.7–1.9 all relate to sediments and sedimentary rocks. The figures have been arranged to introduce the common sediments, the deposition of sediments and the lithification processes that form coastal sedimentary rocks. Similar stories can be told for other sedimentary depositional environments, such as river terraces, deltas, under glaciers and within deserts. Soils (young sediments), which are very important in engineering geology, play only a small part in Figures 1.6–1.9 and are discussed further in Figure 2.1, Parts 3 and 4 and the Appendix.

*Sediments* are mainly derived from the weathering and subsequent erosion of rocks, followed by redeposition on the surface of the Earth. When rocks are exposed to air and water, they slowly break down to form soil, either in situ or after transportation (e.g. by wind, water or ice), and are subsequently deposited. Natural transportation processes are dominated by water, which sorts and selectively deposits its sediment load, usually graded by grain size. Ultimately, nearly all sediments are deposited in the sea as stratified layers or beds of sorted material. They may be temporarily stored en route to the sea as, for example, alluvial river terraces. Loading on the loose and unconsolidated material by subsequent deposits eventually turns the sediments into sedimentary rocks. Other types of sediments are made from organic materials or by chemical processes.

*Sediment transportation.* The most abundant sediment is detrital (or clastic) material consisting of particles of clay, silt, sand and coarser debris, all derived from the weathering and erosion of older rocks. Rivers move the majority of sediments on land, with the coarser debris being rolled along the river bed and finer particles carried in suspension. The sediment is generally sorted into single sizes during transportation (i.e. mainly sands or mainly clays) by decreases in the water velocity. Sediments are also moved in the sea, mainly in coastal waters (Figure 1.6), where wave action and near-shore currents can reach the shallow sea floor. Other transport processes are less dominant, but include gravity, wind, ice and volcanoes. Gravity can result in landslides on slopes, producing materials such as slope colluvium (fine) and steep-slope taluvium (coarse) (see Figure 3.9 and Appendix). Wind can move fine, dry particles up to sand size (see Figure 3.6). Ice is a powerful transporting agent, but is restricted to cold environments (see Figures 3.1 and 3.2). Volcanoes can blast debris over the landscape (see Figure 1.4).

Some minerals are transported by solution in water and are then precipitated in an appropriate chemical environment. The main soluble mineral is calcite (calcium carbonate) and this is widely deposited to form limestone. Evaporite sediments, including chlorides (rock salt) and sulphates (gypsum), are deposited under evaporating conditions. Organic sediments also contain carbonaceous materials; these may be formed in situ (e.g. peats) and also include algal muds and shell debris, which eventually form different varieties of limestone (see Figure 1.9).

*Coastal longshore drift* and onshore/offshore exchanges of coarser sediment are driven by wave energy. Storms arrive at infrequent and irregular intervals, removing material from the upper beach berm and depositing it offshore in the form of bars. Subsequent smaller waves slowly sweep the material back onto the beach. Longshore drift generally occurs in pulses during and immediately after storms. Over the course of a year, waves arrive at a beach from a range of directions and, as the wave approach angle changes, so the local direction of the longshore drift changes.

*Coastal cells.* The shoreline consists of a series of interlinked systems within which sediment is moved along energy gradients from high-energy sources (e.g. eroding cliffs or near-shore sandbanks) by sediment transportation pathways (e.g. beaches) to temporary low-energy sediment stores such as sand dunes or offshore banks. The boundaries of sediment cells occur where there is a zone without breaking waves or where the angle of wave approach is parallel to the beach. These can be fixed, such as at headlands or river mouths, or free, as in energy convergence or divergence zones.

*Beaches.* Figure 1.6 shows a common coastal depositional environment and introduces a typical situation in which local physical and chemical systems control the final deposition of sediments. Other examples of depositional situations include river flood plains, mountain valleys and the snouts of glaciers. Quaternary changes in sea level have played an important part in the history of coastal evolution (see Introduction, Bibliography, Group A books, and Fookes *et al.*, 2005).

Beaches are stores of gravel and sand supplied from source areas on the adjacent coastline and offshore. They can be viewed as part of a larger system, a coastal cell (or sediment transport cell) within which a range of sediment transfers takes place. Beach-building materials might be supplied from the seabed, moved onto the shore by wave energy, or supplied from the land by rivers and eroding cliffs. This material is then redistributed along the shoreline by waves (longshore drift) unless prevented by barriers such as headlands or recent groynes and harbour works. Some of the material can be lost from the cell around the seaward end of such barriers to be later deposited further down the coast or in the deeper sea, particularly during large storms. This can result in a net loss of local beach-building materials. An example of such a loss has occurred from Dawlish beach (Devon, UK), which once protected Brunel's Victorian promenade and railway – these are now subject to storm wave attack.

[above] Eroding cliffs of weak sandstone in a temperate environment; these are a major source of beach sediment.

[right] Weathering cliffs feed an active scree slope that adds to the sediment load carried by the river out to the coast.

Clay, siltstone, sandstone and conglomerate [across the top] are the sedimentary rocks formed from mud, silt, sand and gravel [beneath]; the clay and the mud have dried, and normally are wet and plastic.

[above] These cliffs of weak glacial till on England's east coast retreat at an average of 2 metres per year by repeated small landslides that are a major supply of beach sediment.

[right] Cusps are developed in beach sediment where the waves approach at right angles, as in a sheltered bay.

An arcuate delta is formed of sandy sediment that is deposited by a river more rapidly than it can be carried away by wave action.

A beach stores sand that is eventually carried out to deeper water.

Sandbanks fill an estuary where a river arrives at sea-level.

Longshore drift extends a barrier spit by carrying beach sediment and river alluvium across the mouth of a river.

Coastal sand dunes are formed, and replenished, by beach sand that is blown inland by the wind.

[left] Timber groynes interrupt longshore drift on a sandy beach and a shingle storm beach at its crest.

[right] Barrier islands are transient offshore sand ridges stabilised by plants on a rising, emergent coastline.

A tombolo formed where beach sand accumulated in the lee of an island extends as a spit to connect to the same island.

A Scottish salt marsh at high tide level, formed of sediment colonised by plants.

Swamp is where a future coal seam is formed if later buried by sand.

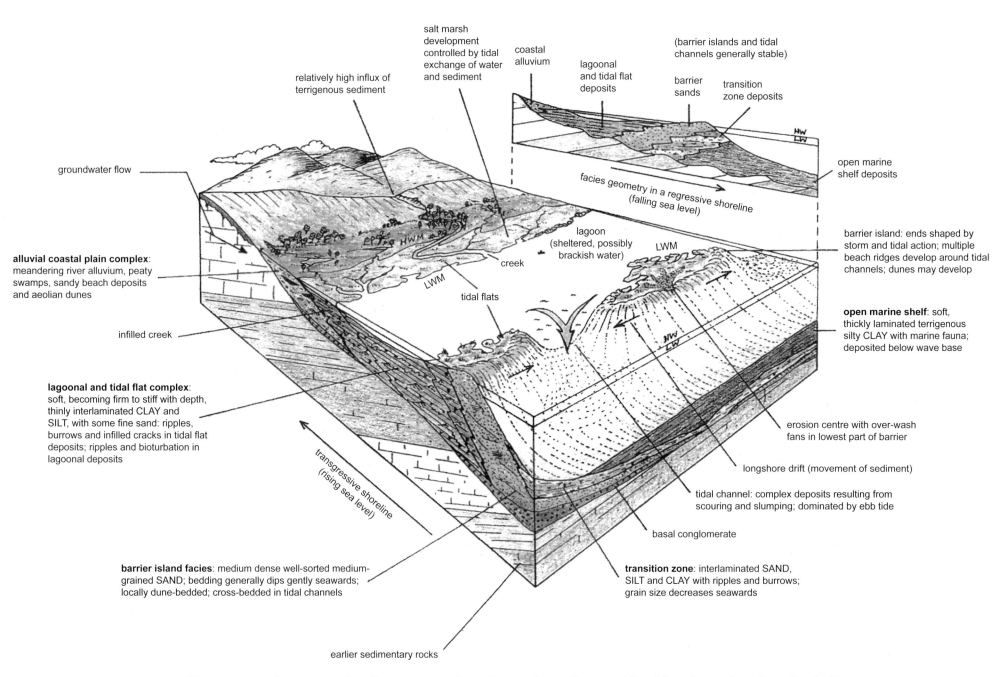

salt marsh development controlled by tidal exchange of water and sediment

coastal alluvium

relatively high influx of terrigenous sediment

lagoonal and tidal flat deposits

(barrier islands and tidal channels generally stable)

barrier sands

transition zone deposits

*facies geometry in a regressive shoreline (falling sea level)*

open marine shelf deposits

groundwater flow

lagoon (sheltered, possibly brackish water)

LWM

creek

tidal flats

LWM

HWM

HW LW

HW LW

**alluvial coastal plain complex**: meandering river alluvium, peaty swamps, sandy beach deposits and aeolian dunes

**barrier island**: ends shaped by storm and tidal action; multiple beach ridges develop around tidal channels; dunes may develop

infilled creek

**open marine shelf**: soft, thickly laminated terrigenous silty CLAY with marine fauna; deposited below wave base

**lagoonal and tidal flat complex**: soft, becoming firm to stiff with depth, thinly interlaminated CLAY and SILT, with some fine sand: ripples, burrows and infilled cracks in tidal flat deposits; ripples and bioturbation in lagoonal deposits

*transgressive shoreline (rising sea level)*

erosion centre with over-wash fans in lowest part of barrier

longshore drift (movement of sediment)

tidal channel: complex deposits resulting from scouring and slumping; dominated by ebb tide

basal conglomerate

**barrier island facies**: medium dense well-sorted medium-grained SAND; bedding generally dips gently seawards; locally dune-bedded; cross-bedded in tidal channels

**transition zone**: interlaminated SAND, SILT and CLAY with ripples and burrows; grain size decreases seawards

earlier sedimentary rocks

*Figure 1.7 Sediments and sedimentary rocks: a linear clastic barrier island beach on the edge of a shelf sea.*

## Sediments and sedimentary rocks: a linear clastic barrier island beach (Figure 1.7)

Shallow shelf sea environments exist on the continental shelf where sediments of cobbles, pebbles and sand of differing grades are deposited, together with various muds and calcareous materials. Their behaviour during storms is of particular significance to coastal engineering.

This figure continues the story started in Figure 1.6 and portrays a hypothetical barrier beach with a mesotidal range of 2–3 m in a temperate climate. The figure models a shallow coastal sea (on a continental shelf) showing the surface erosional and depositional activities and the way in which deposits build up. It gives a fairly specific, but simplified, example of a coastal system (for further information, see Bibliography, Group A books, and Fookes *et al.*, 2007). Figure 1.8 shows what this barrier beach could look like a few tens to hundreds of millions of years in the future when the sediments have been changed (lithified) to their equivalent sedimentary rocks, assuming that there is no disruption due to plate tectonic activity.

*Barrier beaches* are free-standing linear sand or shingle features that tend to run parallel to the shoreline. They may be topped by back-beach deposits, including dunes, and may extend from several hundreds of metres to well over 100 km in length. The barrier beach separates the open shelf environment from tidal flats, lagoons and salt marsh systems, which usually pass landward into coastal plains. Barrier beaches are commonly associated with the following conditions.

- Generally moderate to low wave energy environments that are occasionally exposed to very high-energy events (e.g. hurricanes and tropical storms), which may cause catastrophic changes.

- High sediment availability during the development phase (e.g. from adjacent rivers or eroding cliffs) and micro- to mesotidal conditions (tidal range <4 m).

- Low angle offshore seabed slopes.

Examples of such beaches are the barrier chains on the eastern coast of the USA (e.g. North Carolina), which total over 310 km in length. In the British Isles, Chesil Beach (Dorset) and Slapton Sands (Devon) are good examples.

The forms of barrier beach vary with tidal range.

- *Microtidal barriers* (<2 m) are dominated by waves and tend to be long, narrow and continuous with a limited number of tidal inlets. The inlets may develop wash-over sites during severe storms. Transgressive barriers occur where the beach deposits override any marsh sedimentation. Regressive barriers allow only limited back-barrier sediments to accumulate and marshes are built on these.

- *Mesotidal barriers* (2–4 m) have a mixed-energy regime and tend to be segmented islands cut by frequent tidal inlets (e.g. the East Friesian Islands, Germany). Tidal flows generate ebb and flood tide deltas around the inlets and multiple beach ridges occur at the ends of the islands.

There is a great diversity of barrier forms, although three main types can be identified.

- *Spits* of sand or gravel attached to the mainland at one end, with the far end in open water. These barriers form because there is an energy gradient between the coast and a section of the shoreline (e.g. an embayment) where wave refraction and divergence result in a lower wave energy regime. Like other forms, they extend into deeper water until a point is reached where the sediment supply is less than the rate of erosion (e.g. Farewell Spit, New Zealand is 3 km long). Spits may have various curved forms (e.g. Spurn Head at the mouth of the Humber, UK). Tombolos are sediment links from the mainland to offshore islands, e.g. Chesil Beach, which links the mainland to the Isle of Portland.

- *Welded barriers* are attached to the mainland at both ends, enclosing a lagoon, e.g. Martha's Vineyard (south of Cape Cod, Massachusetts, USA), or wetlands such as Slapton Sands (Devon, UK). They are common on microtidal coasts where there is insufficient tidal energy to maintain a tidal inlet.

- *Barrier islands* (as shown in Figure 1.7) frequently occur in chains parallel to the shore, commonly extending over 100 km, with individual islands separated by tidal inlets, e.g. Padre Island (Texas, USA), which is over 200 km long.

- Back-barrier deposition is dependent on the presence of tidal inlets. *Continuous barriers.* Impeded drainage of fresh water behind the barrier allows the development of lagoons infilled with land deposits, such as peats, e.g. Slapton Ley (Devon, UK).

- *Discontinuous barriers.* These allow the free inflow of salt water into the back-barrier. The sediments are a mixture of marine-derived and terrestrially (land) derived sediments, including over-wash deposits, marine muds, sands and peats (e.g. Chesil Beach and the Fleet Lagoon, UK).

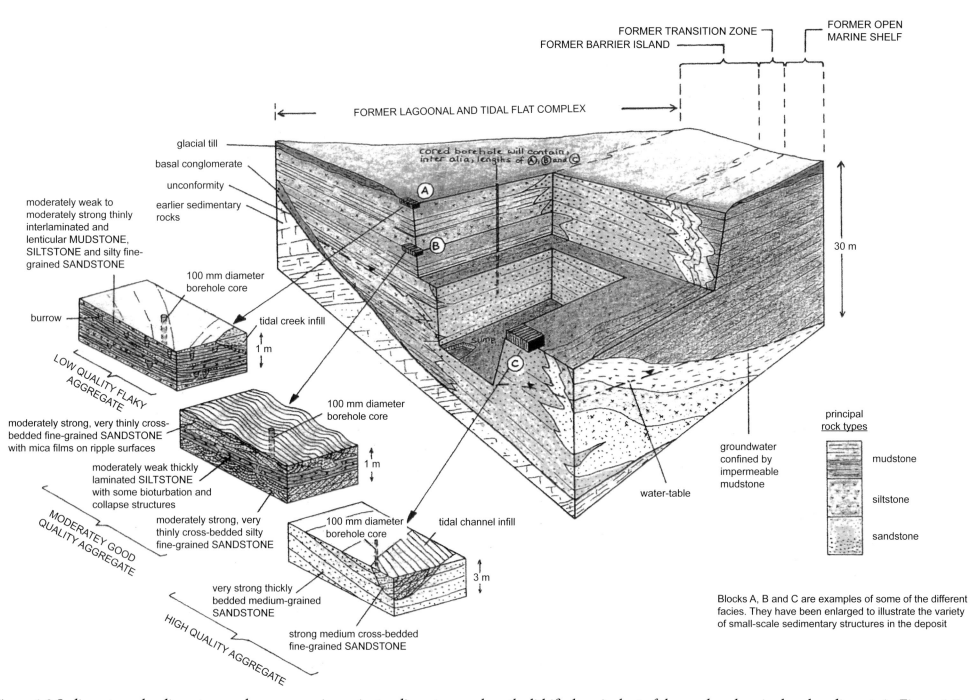

FORMER TRANSITION ZONE

FORMER BARRIER ISLAND

FORMER OPEN MARINE SHELF

FORMER LAGOONAL AND TIDAL FLAT COMPLEX

glacial till

cored borehole will contain, inter alia, lengths of Ⓐ, Ⓑ and Ⓒ

basal conglomerate

unconformity

earlier sedimentary rocks

30 m

moderately weak to moderately strong thinly interlaminated and lenticular MUDSTONE, SILTSTONE and silty fine-grained SANDSTONE

100 mm diameter borehole core

burrow

tidal creek infill

1 m

LOW QUALITY FLAKY AGGREGATE

moderately strong, very thinly cross-bedded fine-grained SANDSTONE with mica films on ripple surfaces

100 mm diameter borehole core

1 m

moderately weak thickly laminated SILTSTONE with some bioturbation and collapse structures

MODERATELY GOOD QUALITY AGGREGATE

moderately strong, very thinly cross-bedded silty fine-grained SANDSTONE

100 mm diameter borehole core

tidal channel infill

3 m

very strong thickly bedded medium-grained SANDSTONE

HIGH QUALITY AGGREGATE

strong medium cross-bedded fine-grained SANDSTONE

groundwater confined by impermeable mudstone

water-table

principal rock types

mudstone

siltstone

sandstone

Blocks A, B and C are examples of some of the different facies. They have been enlarged to illustrate the variety of small-scale sedimentary structures in the deposit

*Figure 1.8 Sediments and sedimentary rocks: a quarry in ancient sedimentary rocks – the lithified equivalent of the modern barrier beach sediments in Figure 1.7.*

## *Sediments and sedimentary rocks: a quarry in ancient rocks – the lithified equivalent of modern barrier beach sediments (Figure 1.8)*

Understanding the lithification of deposited sediments helps in the production of the initial model of the engineering site, in part because the extent of lithification is a major factor in determining the strength and mass of the rocks. Induration is the process of hardening of sedimentary rocks over time.

Figure 1.8 models the lithification (the transformation of sediments into rocks) of the modern barrier beach shown in Figure 1.7 over a moderately long period of geological time (commonly at least millions to tens of millions of years). The figure is again hypothetical and illustrates what may happen to modern sediments deposited in a geological system. Only a small proportion of geological deposits are preserved as hard rocks, and even less on land than in marine situations. Preservation depends on whether the burial conditions are right; if post-depositional erosion occurs, then the sediments will be recycled (Bibliography, Group A books, and Selley, 1996).

In general, only some of the sediments in a depositional environment will be preserved. Subsequent diagenetic (low temperature and pressure) changes and new near-surface conditions influence the lithification processes, so it is usually difficult for geologists to interpret the original depositional conditions. The geological cycle ensures that near-surface rocks will eventually be modified by the heat and pressure of tectonic forces, often completely changing their characteristics so that the original sediments may not be recognized. The inevitable fate (typically over a very long period of geological time) of all crustal rocks is to become weathered and removed by erosion, which makes them available for a new cycle of deposition.

### LITHIFICATION

There are two principal groups of sediments: clastic (particles or grains) and non-clastic (produced by chemical or biochemical precipitation). Clastic sediments are composed of fine (clay-sized) to much coarser grains. Gravel is a typical coarse-grained clastic sediment. All sedimentary rocks are layered, although the layering may not be recognizable in a small exposure of thick-bedded (up to metres between bedding planes) or 'massive' (without discontinuities) sedimentary rock.

For unconsolidated (i.e. as-deposited) sediments to be turned into a stronger sedimentary rock, they must be lithified. Several lithification processes take place over short to long periods of geological time depending on the conditions and the original sediment type. The texture of a sedimentary rock refers to the shape, size and arrangement of its constituent particles, all of which can be described and measured in the field or laboratory. The matrix of a sedimentary rock refers to the fine material trapped in the pore spaces between coarser grains. Fine-grained sediments have a higher porosity than those with coarser grains and can therefore undergo a greater amount of consolidation. Muds and clays may have original porosities of up to about 80%, compared with up to about 50% in sands and silts. The amount of volume lost on consolidation in sands and silts is usually low, but can be as high as 25%; in clays it may be up to 55% (see Appendix for further discussion).

The consolidation of clays and silty clays generally refers to the increase in strength resulting from the restructuring of particles to a denser configuration (individual particles becoming closely packed and even deformed), the loss of water (originally trapped as the sediment was deposited) and reduced porosity. These processes are all caused by compaction under load. It can also include some cementation and new mineral growth. *Normally consolidated* clays are those which have never been under a higher load than that of their existing overburden. Most young soils are in this condition. However, clays that have been under a higher load in their geological past (buried by sediments deposited later) become *overconsolidated* when some or all of the covering rock is subsequently removed by erosion. Most bedrock clays fall into this category (e.g. mid-Tertiary age London Clay). Some clays have become overconsolidated as a result of previous ice-loading or as a consequence of former horizontal tectonic stress. Such clays have a lower porosity and higher strength as a result of the overconsolidated process and become 'stiff'. Ultimately, claystones (very stiff clays with some cementation) and laminated shales may be formed under very high loads (see Appendix for engineering considerations related to overconsolidated).

Sediments coarser than clays are composed of granular material, mostly quartz, although mica, feldspar, rock fragments and volcanic debris are relatively common. Organic debris is typically dominated by calcite (calcium carbonate) from marine or freshwater shell debris and forms a special case. It recrystallizes to form limestone and related rocks.

The following processes are common in lithification.

- *Compaction.* Restructuring to a tighter grain packing with a decrease in the inter-grain volume as a result of the burial load. An increase in strength occurs due to closer grain-to-grain packing and the loss of water.

- *Cementation.* The filling of inter-granular pore spaces by deposits of a mineral cement (e.g. calcite) brought in by circulating groundwater or derived from the partial solution of the grains. A cement may subsequently be fully or partially removed by leaching. The rock strength and colour are dependent on the type or types of cement. Silica is usually the strongest cement, followed by iron oxides and calcite, with clay the weakest. Siltstones, sandstones and conglomerates are typically lithified by cementation.

- *Recrystallization.* The small-scale solution and redeposition of minerals. As some grains become smaller and others become larger, a closer packing arrangement may produce a stronger mosaic texture. This is the dominant process in carbonate rocks and accounts for the great strength of many limestones. Metamorphism is an extreme state of alteration involving high temperatures and/or pressures and results in a much greater degree of recrystallization. It is not considered to be a sedimentary lithification process (see Figure 1.5).

Coal Measures are sequences of sandstones, commonly buff coloured, and shales (which are generally grey and darker than the sandstones) with about 2% of thin black coal seams and thin grey seatearths.

[above] Alternating sequence of thinly bedded sandstones, shales and siltstones uplifted tectonically but still almost horizontal.

[below] Horizontal flagstone, sandstone that breaks into thin flags, in an old mine, with a roof support pillar made of waste flagstones.

[above] Strong, coarse sandstone may be known as gritstone, and was widely cut into millstones.

[above right] This cross-bedded sandstone is a lithified sand dune about 15 metres tall.

[right] Cross bedded sandstone that was formed on the front of an advancing river delta.

Three sandstones of varying strength, all exposed and eroded in semi-arid environments: [left], a poorly consolidated, weak sandstone that is easily eroded into badlands; [middle], a medium strong sandstone that has been carved to create a giant statue of Buddha; [right], a strong sandstone that stands in vertical cliffs hundreds of metres tall along a canyon.

Crags of coarse, strong sandstone, known as gritstone, with weathered cliff profiles that reflect variations in the mineral cementation and erosional resistance of the rock.

[above] Dipping sandstones in beds of different thickness, all degrading to form a footslope debris of blocks and loose sand.

[right] Two mudrocks of varying strength: above, a weak and poorly lithified clay that is easily eroded into badland terrain; below, a strong, bedded mudstone that has been well lithified and folded before erosion into rock slabs.

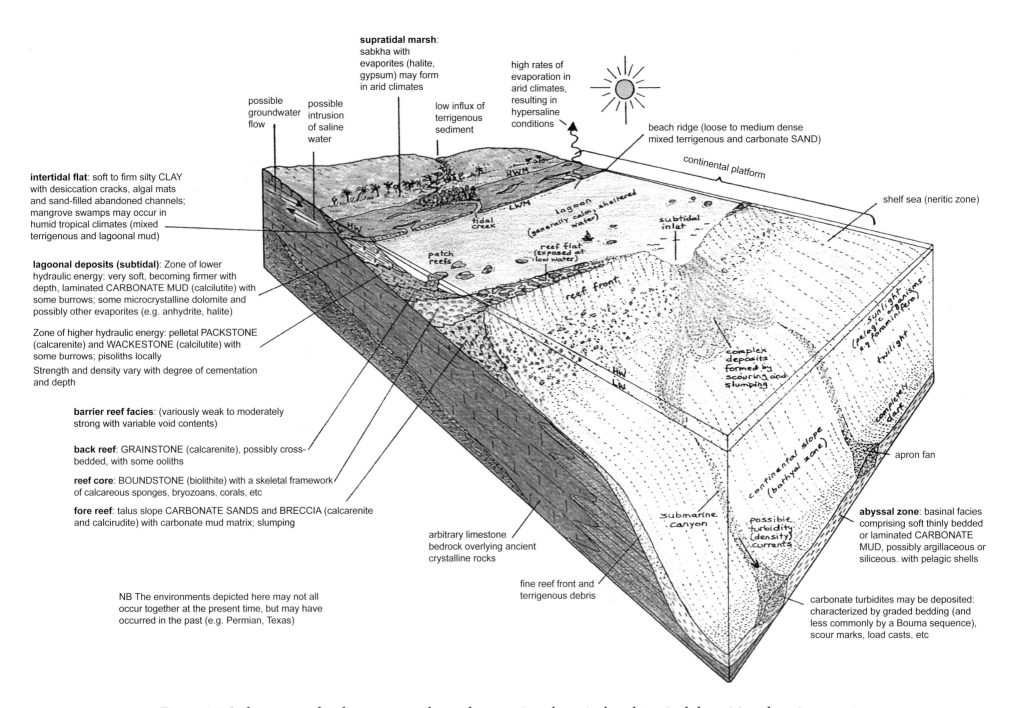

*Figure 1.9 Sediments and sedimentary rocks: carbonates in subtropical and tropical depositional environments.*

## *Sediments and sedimentary rocks: carbonates (Figure 1.9)*

Figure 1.9 is an idealized pictorial block model of subtropical and tropical carbonate marine shelf *facies* (closely related rock types) and illustrates some of the most common environments that produce mainly carbonate sediments and, eventually, limestone rocks. This figure introduces some of the characteristics of the extensive and complex carbonate family (for more detailed discussion, see Bibliography, Group A books, and Bell, 2000).

### CARBONATE ROCKS

By definition, carbonate sediments and rocks contain more than 50% carbonate minerals (minerals composed of carbonate anions combined with one or more cations). The most common carbonate mineral is calcite (calcium carbonate, $CaCO_3$), which is the principal component of limestones. The term carbonate can be used to describe all rocks and sediments composed principally of carbonates, including dolomites, dolostones and the other dolomitic rocks that contain high proportions of magnesium carbonate. Iron and other less common carbonates can also be included. For basic clarifications of mixed carbonate rocks see Tables 1.9.1 and 1.9.2.

Carbonate sediments are formed in many marine and some terrestrial depositional environments, but are most abundant on the floor of shallow tropical seas. Carbonate minerals precipitate from carbonate-saturated water by biochemical or chemical processes and accumulate in many ways: as the skeletal remains of organisms; as inorganic growths that nucleate on fine mobile debris in shallow seas; as crusts within arid soils; as laminated precipitates on the walls of limestone caves; and on the ground around hot-springs. When a body of water contains salts, these crystallize out as the water evaporates. The first salts to be deposited are carbonates, which are less soluble than the other common evaporite minerals (e.g. sulphates, chlorides) deposited subsequently. Typical locations of crystal deposition are narrow arms of the sea in the tropics, land-locked basins and hot desert lakes. The bodies of water currently depositing evaporites (e.g. the Dead Sea, the Caspian Sea and parts of the Arabian Gulf, including local sabkhas in coastal environments) are relatively minor compared with the numerous ancient evaporating seas that occurred in earlier, hotter periods of the Earth's history (e.g. the Permian). Today, there is insufficient evaporation of salt-containing waters for any minerals other than carbonates to be precipitated.

All carbonate sediments have a similar basic chemistry, but the processes by which they are formed vary. Carbonate sediments include the following types.

- Material precipitated from water as a result of evaporation and the concentration of the remaining solution, which eventually forms a rock or cement of chemical origin.

- Particulate (clastic) materials, such as fragments of older limestones that were transported and eventually cemented to form a rock of mechanical origin (e.g. conglomerates and breccias).

- The skeletal remains of animal bodies made of carbonates, such as broken shells, coral reefs and similar organic material, which eventually become cemented to form a rock of mainly biological origin.

- The widespread white/cream rock of North America, the Middle East and Europe, 'chalk', is predominately composed of tiny deep-sea calcareous organisms (e.g. coccolithophores and foraminifera), which fall to the sea floor after death.

The engineering characteristics of young carbonate sediments are related to their texture. The texture is directly related to the turbulence of the water in which the sediment was deposited. In a particular sediment, the relative proportion of mud matrix to cement is an index of the water turbulence or mechanical energy. Depositional areas with higher energies produce clean, well-sorted, coarse-grained carbonate sediments, whereas areas of lower energy are responsible for muddy deposits. High-energy sites include beaches, surf zones, dunes and tidal channels. Fine-grained carbonate sediments, i.e. muds, generally accumulate in calmer protected areas such as lakes, lagoons, deep-sea basins and areas on the lee side of major islands on oceanic banks. The term 'mud' is preferred to 'clay' (for clay-sized particles) in fine carbonate sediments as these are primarily composed of fine carbonate particles and not clay minerals; 'lime mud' can be used for emphasis. Low-energy depositional sites typically produce strong, dense limestones, although this situation can be complicated by subsequent recrystallization.

The engineering properties of older carbonate sediments are influenced by the grain size and post-depositional changes that bring about induration (hardening) and increase the strength of the rocks. Limestone is perhaps more prone to pre- and post-consolidation changes than any other rock type. After burial, limestones can be modified by chemical changes to such an extent that their original characteristics are obscured or even obliterated. The induration of carbonate sediments often starts during the early stages of deposition as a result of cementation, which occurs where individual grains are in contact (e.g. many of the Tertiary limestones in Britain). Cementation is therefore not solely dependent on the influence of denser packing resulting from an increase in overburden pressure. As induration can take place at the same time as deposition, this means that carbonate sediments can sustain high overburden pressures so that they become moderately strong, yet still retain high porosities at considerable depths (e.g. many Jurassic limestones in Britain). A layer of cemented grains may even overlie one that is poorly cemented. High overburden pressures, creep and recrystallization eventually produce a crystalline limestone with a very low porosity and very high strength (e.g. Carboniferous limestones in Britain).

In summary, the engineering performance of a limestone cannot be judged only on its name as there are many variables that influence the engineering characteristics of the limestone family. These include the following variables.

- *Mineral composition.* The different carbonate minerals include: calcite ($CaCO_3$), the most common carbonate material, which, on recrystallization, produces strong rocks; aragonite ($CaCO_3$), which only occurs in younger, weaker limestones; dolomite ($CaMg(CO_3)_2$), which typically produces a stronger stone valued as an aggregate source; and siderite ($FeCO_3$), which is an ironstone and is less common. Carbonates may be admixed with clay minerals or quartz.

- *Origin*. This influences the rock strength. Carbonates may be formed from shallow detrital (clastic) sediments, by chemical precipitation, by reef-building organisms, or from deep-sea micro-fossils (e.g. chalk).
- *Grain size and texture*. These can both vary enormously.
- *Degree of induration*. Recrystallization and the degree of induration (if any) both largely account for the dry rock strength, which can range from weak to very strong. Note that the saturated strength is typically a little less than the dry strength.

At any one site, limestones may grade horizontally and/ or vertically into related sedimentary types – for example, a pure limestone with an increasing clay content may grade into a limey marlstone, then a marlstone, clayey marlstone and, finally, a pure claystone. A marl is a mixture of clay and calcium carbonate (Tables 1.9.1 and 1.9.2).

Table 1.9.1 Classification of impure carbonate rocks (after Fookes, 1988).

*Silt, sand and gravel carbonate rocks*

*Clay carbonate rocks*

Table 1.9.2 Classification of mixed carbonate rocks in percentage carbonate and predominant grain size (after Dearman, 1981).

Note: (1) Non-carbonate constituents are rock fragments or quartz, micas, clay minerals (2) Predominant grain size implies over 50%.

The environments of limestone formation in tropical seas around Fiji: [above], a lagoon inside a coral reef, and [below], a reef-fringed shelf with a central island that is essentially a sand mound of coral fragments.

Reef environments in warm waters where many types of limestone originate: [above] numerous types of hard coral form the calcite framework of a reef; [above right] corals are just one part of the complex life assemblage forming a reef in shallow water that is clear and clean because it is free of any detrital sand or mud derived from land.

A beach of white sand formed entirely of fragments of shell washed in by the waves from the sea bed.

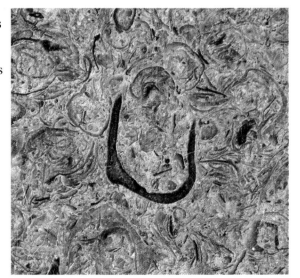

[above] Polished face of a slab of limestone formed almost entirely of shell fragments (the large dark shell is 10 mm across).

[left] A lagoon beach that is composed entirely of the shells of a single species of bivalve cockle.

Chalk cliffs 70m high along the English coast; though the chalk is a weak rock, the cliffs are kept vertical by rockfalls above marine toe erosion that is faster than surface erosion where rainwater sinks underground.

[left] A giant salt pan on the Andean Altiplano; it becomes a lake every winter, when new salt crystals grow along desiccation cracks that formed in the summer.

[below] A lens of strong, massive, unbedded limestone, about 30 metres thick, originated as an algal reef within a bedded sequence of tropical lagoonal limestones.

A weak, chalky limestone is sawn by hand into blocks that case-harden by re-deposition of carbonate in the surface layers after a few weeks of exposure and drying.

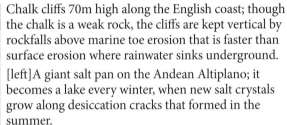

Thin bands of limestones interbedded with shales form the distinctive rock sequences of the Lias in England.

Horizontal, strong, bedded limestone forms the vertical cliffs that rise more than 100 metres on each side of a narrow gorge lost from sight in deep shadow.

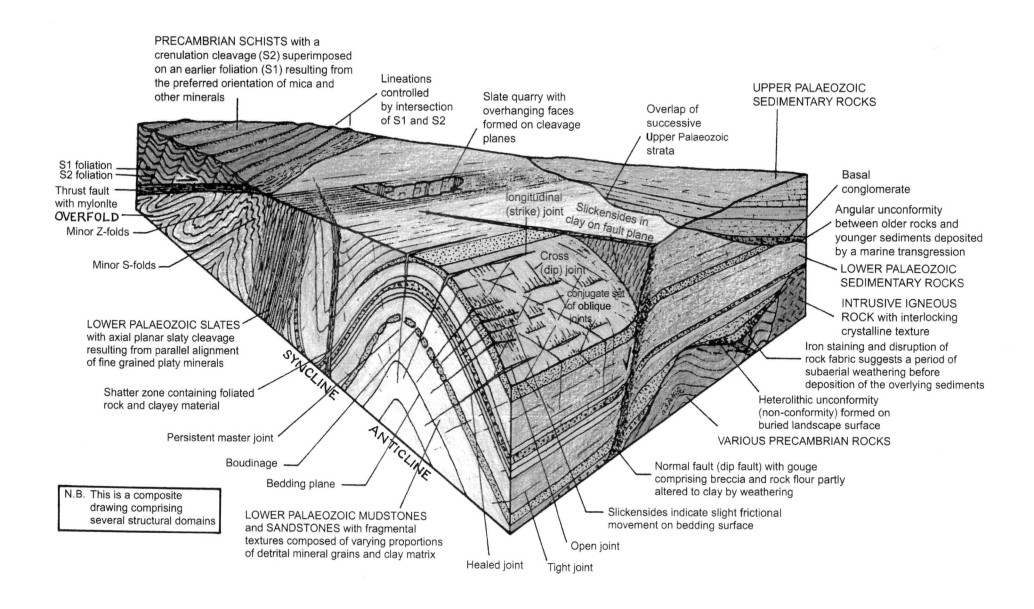

PRECAMBRIAN SCHISTS with a crenulation cleavage (S2) superimposed on an earlier foliation (S1) resulting from the preferred orientation of mica and other minerals

Lineations controlled by intersection of S1 and S2

Slate quarry with overhanging faces formed on cleavage planes

Overlap of successive Upper Palaeozoic strata

UPPER PALAEOZOIC SEDIMENTARY ROCKS

S1 foliation
S2 foliation

Thrust fault with mylonite
OVERFOLD

Minor Z-folds

Minor S-folds

longitudinal (strike) joint

Slickensides in clay on fault plane

Basal conglomerate

Angular unconformity between older rocks and younger sediments deposited by a marine transgression

LOWER PALAEOZOIC SEDIMENTARY ROCKS

Cross (dip) joint

conjugate set of oblique joints

INTRUSIVE IGNEOUS ROCK with interlocking crystalline texture

LOWER PALAEOZOIC SLATES with axial planar slaty cleavage resulting from parallel alignment of fine grained platy minerals

Iron staining and disruption of rock fabric suggests a period of subaerial weathering before deposition of the overlying sediments

Shatter zone containing foliated rock and clayey material

SYNCLINE

Heterolithic unconformity (non-conformity) formed on buried landscape surface

VARIOUS PRECAMBRIAN ROCKS

Persistent master joint

ANTICLINE

Boudinage

Bedding plane

Normal fault (dip fault) with gouge comprising breccia and rock flour partly altered to clay by weathering

N.B. This is a composite drawing comprising several structural domains

LOWER PALAEOZOIC MUDSTONES and SANDSTONES with fragmental textures composed of varying proportions of detrital mineral grains and clay matrix

Slickensides indicate slight frictional movement on bedding surface

Open joint

Healed joint
Tight joint

*Figure 1.10 Geological structures (after Fookes 1997a).*

## Geological structures (Figure 1.10)

The stresses generated by plate tectonics can be accommodated by extremely slow plastic flow and the deformation of rocks, or by fracturing. The structure of the rock and, consequently, the rock mass is related to the regional and local geological stress histories. An understanding of geological structures can help to unravel the local or site-specific structural conditions and can be incorporated into the geomodel. The term 'strata', as used here, is applied to rocks that form layers or beds. For further discussion, see Bibliography, Group A books.

### DEFINITIONS

The two dominant groups of features produced when strata are deformed by movements of the Earth's crust are *folds*, where the rocks have become buckled, and *faults*, where the rocks fracture. A fold is produced when a more or less planar surface or rock unit is deformed to give a curved surface or feature. A fault represents a surface of discontinuity along which the strata on either side have been displaced relative to each other.

The orientation of beds, bedding planes, folds, faults and any other planar structures are described in geology by their 'strike' and 'dip'. The strike is the direction in which a horizontal line can be drawn on a dipping (inclined) planar surface. This term is also used in the same sense to indicate the general trend or run of the beds in an area. The true dip of a plane is the angle that it makes below the horizontal, the angle being measured in a direction perpendicular to the strike. The apparent dip is the angle measured in any other direction, usually on small, poorly exposed locations in the field.

The hade is the angle between a fault plane and the vertical. The hanging wall refers to the fault block that lies above any inclined fault surface, whereas the foot wall refers to the fault block that lies below any inclined fault surface. The vertical component of displacement along a fault plane is called the throw and the heave refers to the horizontal displacement. Where the displacement along a fault includes a vertical component, the down-throw and up-throw refer to the relative movements of strata on opposite sides of the fault plane. These terms are in common usage, especially in underground mining.

Non-tectonic (diagenetic) structures are formed in sediments during diagenesis changes at low temperatures and pressures after deposition and can include small-scale folding, jointing, faulting, fissuring and lithification. These features are particularly important in the behaviour of engineering soils.

### COMMON STRUCTURAL FEATURES

*Joints* are minor tensile or very minor shear fractures within a rock mass along which no significant movement has occurred. They can develop in nearly all rocks. The frequency or density of joints and their length can be extremely variable and these factors are the primary features that determine the strength of a rock mass. Joints commonly occur in distinct patterns related to the history of the region and can form sets of parallel joints, or joint systems involving several sets. Planes of weakness in foliated metamorphic rocks develop independent of the bedding, causing rock cleavage and schistosity (see Figure 1.5).

In the UK, the word 'discontinuity' is used in engineering for all forms of fracture, including joints; however, in the USA and Australia the word 'defect' is used. Discontinuities, including bedding, tend to be the dominant control in the stability of rock slopes. A geomechanical survey of the discontinuities in a rock slope is an important component in the evaluation of hazards from rock-falls and in pre- or post-slope failure construction work on rock slopes.

*Faults* are discontinuities (fractures) on which movement between the opposite faces of the fault has occurred (Figure 1.10).

- *Normal faults* (or extensional faults) are those in which the hanging wall has moved downwards relative to the foot wall. Ground ruptures can occur along multiple fault splays (series of small branching faults) across zones several kilometres wide.

- *Reverse faults* are those in which the hanging wall has moved up relative to the foot wall. Fault traces typically form a sinuous, discontinuous line across the ground surface. Such fault movements commonly cause earthquakes associated with compressional stresses. They can be repeated over geological time and accumulate to form a broad zone of ground deformation up to hundreds of metres wide.

- *Strike-slip faults* (transcurrent wrench or tear faults) are those in which movement is essentially horizontal parallel to the fault strike. The faults are either right-lateral (dextral) or left-lateral (sinistral) depending on the relative motion of the block on the opposite side of the fault. In earthquakes, active surface ruptures, which are almost entirely secondary, are the results of near-surface deformation during earthquake vibrations. They can cause a zone, many kilometres wide, of ground-cracking and the bulging and tearing of near-surface materials. The transform faults that form some plate boundaries are a type of strike-slip fault.

Sustained stress under high confining pressures below the ground surface can lead to a range of folded bedrock structures, from gentle tilting and doming to intense deformation and crumpling (see Figure 1.10).

Most *folds* originate at depth and are formed slowly in geological time. Simple forms include *anticlines* (up-folds with the oldest beds in the centre of the fold structure), *synclines* (down-folds with the youngest beds in the centre) and *periclines* (dipping anticlines). These folds are symmetrical if the dips on the opposing flanks are the same, otherwise they are asymmetrical (e.g. over-folds, recumbent folds). *Monoclines* are one-limbed folds; on either side of monoclines the strata are horizontal or only gently dipping. Periclines are anticlines that are folded so that they appear as very elongate domes in plan. Recumbent folds occur where rocks are overturned and are the product of major horizontal compression; this may lead to shearing in the upper part of the fold so that a nappe is created above a thrust fault (a type of low angle reverse fault).

Folds near the ground surface and exposed by erosion can dominate the landscape with forms that mimic the fold pattern (e.g. the Zagros Mountains, Iran). Inverted relief occurs where long-term weathering and erosion have preferentially removed the weaker or more fractured beds in the core or flanks of the fold, leaving a valley along an anticline or a ridge along a syncline.

[above] Angular anticline in limestone on a foreshore.

[below] Rounded anticline in sandstone in a riverbank.

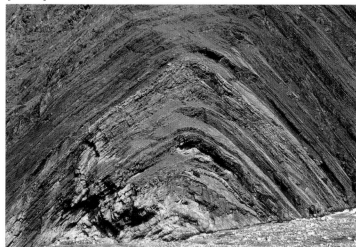

[above] Anticline and syncline in a steeply folded sequence of interbedded limestones and shales.

[right] Zig-zag of angular, recumbent folds in a greywacke sequence, with a person for scale.

[below right] Half a metre of beds within a folded evaporite sequence of gypsum and limestone .

[below] A glacial drag fold exposed in a sea cliff, and formed where Pleistocene ice flowed from the right over the thinly bedded chalk so that it rucked up the upper layers of the frozen ground.

[left] A mountain formed by a limestone anticline.

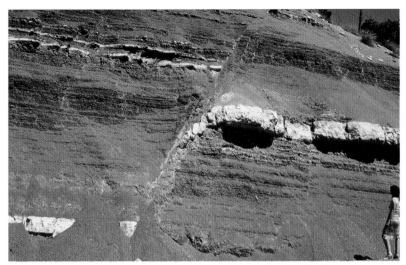

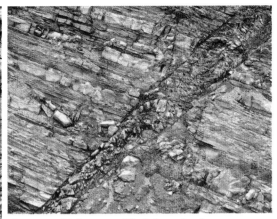

A metre-thick zone of fault breccia along an inclined fault in dipping shale and sandstone.

A normal fault with a shale sequence containing bands of white limestone, all displaced down to the left by about two metres.

A reverse fault in limestones, with a strong metre-thick bed displaced to the left above the fault; thinner beds have accommodated the compression by folding.

[right] Variation in spacing of vertical joints in a limestone.

[below] A rockslide due to bedding-plane failure in dipping limestone. The location of the car would be unsuitable in wet weather when the next slide could occur.

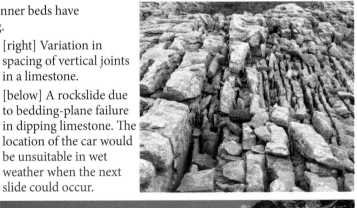

[left] A major unconformity with horizontal limestone above steeply dipping greywacke. [right] Vertical joints in massive sandstone, with rockbolts placed to keep the rock tight, dry and stable in a road cutting.

Banded shales in tension, with a central block that has dropped down between two normal faults, exposed in a road cutting.

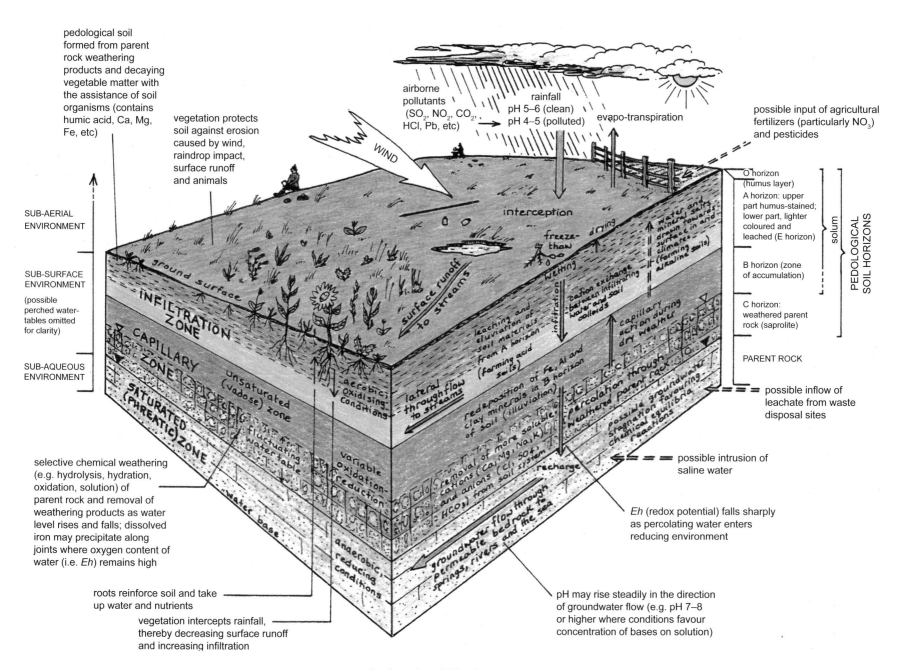

pedological soil formed from parent rock weathering products and decaying vegetable matter with the assistance of soil organisms (contains humic acid, Ca, Mg, Fe, etc)

vegetation protects soil against erosion caused by wind, raindrop impact, surface runoff and animals

airborne pollutants ($SO_2$, $NO_2$, $CO_2$, HCl, Pb, etc)

rainfall pH 5–6 (clean) pH 4–5 (polluted)

evapo-transpiration

possible input of agricultural fertilizers (particularly $NO_3$) and pesticides

WIND

interception

SUB-AERIAL ENVIRONMENT

SUB-SURFACE ENVIRONMENT

(possible perched water-tables omitted for clarity)

SUB-AQUEOUS ENVIRONMENT

ground surface

INFILTRATION ZONE

CAPILLARY ZONE

SATURATED (PHREATIC) ZONE

unsaturated (vadose) zone

fluctuating water table

aerobic: Oxidising Conditions

variable oxidation-reduction

anaerobic, reducing conditions

water base

freeze-thaw

wetting

drying

surface runoff to streams

leaching and eluviation of soil materials from A horizon (forming acid soils)

infiltration

lateral throughflow to streams

redeposition of Fe, Al and clay minerals in B horizon of soil (illuviation)

removal of more soluble cations (Ca, Mg, Na, K) and anions (Cl, $SO_4$ + $HCO_3$) from soil system

cation exchange between infiltrating water and soil colloids

capillary action during dry weather

percolation through weathered parent rock

possible groundwater stagnation favouring chemical equilibria reactions

recharge

groundwater flow through permeable bedrock to springs, rivers and the sea

water and mineral salts drawn towards surface in arid climates (forming saline, alkaline soils)

O horizon (humus layer)

A horizon: upper part humus-stained; lower part, lighter coloured and leached (E horizon)

B horizon (zone of accumulation)

C horizon: weathered parent rock (saprolite)

solum

PEDOLOGICAL SOIL HORIZONS

PARENT ROCK

possible inflow of leachate from waste disposal sites

possible intrusion of saline water

Eh (redox potential) falls sharply as percolating water enters reducing environment

selective chemical weathering (e.g. hydrolysis, hydration, oxidation, solution) of parent rock and removal of weathering products as water level rises and falls; dissolved iron may precipitate along joints where oxygen content of water (i.e. Eh) remains high

roots reinforce soil and take up water and nutrients

vegetation intercepts rainfall, thereby decreasing surface runoff and increasing infiltration

pH may rise steadily in the direction of groundwater flow (e.g. pH 7–8 or higher where conditions favour concentration of bases on solution)

*Figure 2.1 Idealized soil block model: temperate zone.*

48

# Part 2. Near-surface ground changes

## *Basic soils and landscapes (Figure 2.1)*

### COMMON SURFACE CONDITIONS

Ground materials consist of unweathered rock (bedrock), which may be overlain by in situ weathered rock (*saprolite*) and/or soils. Engineers describe any non-lithified soils that overlie solid rock as *overburden,* although this is known to geologists and geomorphologists as *regolith*. Regolith may consist of saprolite, alluvium, glacial till, wind-blown loess or dune sand, volcanic dust and various other unconsolidated materials.

The nature of the bedrock underlying an area is the product of its geological history. This includes the mode of deposition of the sedimentary rocks forming part or all of the bedrock and any post-depositional (diagenetic) changes in the sediments. These changes include compaction, lithification, cementation and weathering changes related to the history of the area (see Figure 1.8). Tectonic activity (folding, faulting and the emplacement of igneous and metamorphic rocks) is not considered to be part of the diagenetic changes. Igneous and metamorphic rocks may also form part or all of the local bedrock.

The extent to which the bedrock geology of an area is reflected in the landscape depends on whether the rocks are covered by a significant thickness of regolith. The enormous range of combinations of rock types, rock structures, weathering and erosional history means that every terrain model constructed during the early part of the site investigation is unique. In practical terms the observed materials within a system or in the ground are not necessarily predictable, so further investigations by boreholes, pits and geophysical surveys are required (see Parts 4 and 5).

In the longer term (e.g. hundreds to tens of thousands of years), the rate and nature of local landscape changes are dependent on the mass characteristics of the rocks and soils, mainly their intact strength, discontinuities and susceptibility to the local weathering and erosion processes of past and present climates.

Earth surface systems can be used to describe how the transfer of sediments and energy (e.g. down-slope creep, erosion along stream channels) produces the relationships between the landforms in an area (Fookes *et al.*, 2007). Local surface systems are primarily controlled by the geological setting, the geographical location (including the climate zone; see Figure 2.2) and the local ground materials. The geological character of the local bedrock and its structure determine the broad-scale form of the local geomorphological landscape and its systems, relief and slope gradients. It also determines the materials available for erosion and transport by surface processes.

Engineering soils are typically described in engineering terms according to their dominant particle size using, for example, the Unified Soil Classification System (Norbury, 2010; Bibliography, Group B books). Three main soil types are recognized.

- *Residual soils* are the product of the in situ weathering of bedrock where the soil thickness and type are broadly associated with the past and present climate and the intensity of weathering (Figures 2.2–2.5, 3.6 and 3.8). A weathered rock profile is created over fresh bedrock and distinctive zones (layers or horizons) can develop in response to variations in the intensity of weathering and the movement of moisture and minerals. The upper layers contain rock debris that has been completely weathered to a soil. Lower down the profile there are increasing amounts of unweathered and partly weathered rock (Figure 2.1). *Tropical residual soils* are a special case of residual soil found in wet tropical areas. They exhibit distinctive engineering properties and characteristics (Fookes, 1997b), ranging in grade VI weathering (see later) from fersiallitic to ferruginous to ferrallitic soils formed by the increasing length of time of weathering and the climate of the area.

- *Transported soils* (e.g. alluvium, loess) are the products of the erosion of residual soils or bedrock that have been transported and deposited elsewhere.

- *Organic soils* are formed in situ by the growth and decay of vegetation – for example, peat, which forms in anaerobic conditions when the ground is waterlogged. Peat formation is encouraged in areas of high rainfall and low temperatures, stimulating further water-logging. About 15% of Ireland is covered by blanket peat bog.

Three main weathering phases are recognized in engineering geology, starting from grade VI at ground level.

- Weathering grades IV to VI (solum or 'true' soil to saprolite or chemically weathered rock), which form a continuum of specific mineral soil development in increasing order of weathering (see Figures 2.4 and 2.5).

- Grades II and III are increasing degrees of weathering developed on the underlying fresh bedrock. Corestones are commonly present in grade III weathering, depending on the original joint spacing.

- Grade I is unweathered fresh bedrock.

It is important to note that some ancient over-consolidated bedrock clays – for example, the Tertiary London Clay and the Jurassic Oxford Clay in Britain, which are not lithified – are likely to be called 'engineering clays' by engineers.

### GROUNDWATER

Typical near-surface groundwater conditions are illustrated in Figure 2.1, which is based on a temperate zone climate. Permeability characteristics are given in Table 2.1.1 and 2.1.2. The rates of evapotranspiration and infiltration generally vary seasonally and from year to year. They may also vary on shorter timescales, perhaps in response to local major storms. It should be noted that conditions during the construction phase may differ significantly from those found during the ground investigation.

Table 2.1.1 Typical permeability values.

| Soil types | Homogeneous clays below the zone of weathering | Silts, fine sands, silty sands, glacial till, stratified clays | Clean sands, sand and gravel mixtures | Clean gravels |
|---|---|---|---|---|
| | | Fissured and weathered clays and clays modified by the effects of vegetation | | |

Coefficient of permeability (log scale)

m/sec
$10^{-11}$  $10^{-10}$  $10^{-9}$  $10^{-8}$  $10^{-7}$  $10^{-6}$  $10^{-5}$  $10^{-4}$  $10^{-3}$  $10^{-2}$  $10^{-1}$  1

cm/sec
$10^{-9}$  $10^{-8}$  $10^{-7}$  $10^{-6}$  $10^{-5}$  $10^{-4}$  $10^{-3}$  $10^{-2}$  $10^{-1}$  1  10  100

ft/sec
$10^{-10}$  $10^{-9}$  $10^{-8}$  $10^{-7}$  $10^{-6}$  $10^{-5}$  $10^{-4}$  $10^{-3}$  $10^{-2}$  $10^{-1}$  1

| Practically Impermeable | Very low | Low | Medium | High |
|---|---|---|---|---|

| Drainage conditions | practically impermeable | Poor | Good |
|---|---|---|---|

Estimation of coefficient of permeability: for granular soils, the coefficient of permeability can be estimated using Hazen's formula:

$$k = c_1 D^2_{10}$$

where $k$ is the coefficient of permeability in m/s, $D^{10}$ is the effective particle size in mm, and $c_1$ is a factor varying between 100 and 150.

Table 2.1.2 Typical ranges of coefficient of permeability (k) for different types and conditions of rock.

| $k$ (m/s) | 1 | $10^{-2}$ | $10^{-4}$ | $10^{-6}$ | $10^{-8}$ | $10^{-10}$ | $10^{-12}$ |
|---|---|---|---|---|---|---|---|
| Clays | | | ← Stratified — — — — —  — Homogeneous — | | | | |
| Shale | | | ← Mass — — — — — — — — — — | | | | |
| Sandstone | | ← — Fractured — — —— — Intact — | | | | | |
| Limestone | ← Solution cavities — — — — — — — — Intact — | | | | | | |
| Salt | | | | | ← — — Bedded — — — — — | | |
| Volcanics | | ← — —Weathered — — — — —— — Intact — | | | | | |
| Metamorphics | | ← — Weathered — — — — — — — — Intact — | | | | | |
| Granites | | ← — — Weathered —— — — — — — — — Intact — | | | | | |

- Where the annual infiltration exceeds the evapotranspiration, groundwater flows downwards from the surface. It eventually reaches a zone of saturation within the capillary fringe, where it is held within the soil pores by surface tension. Fine-grained soils within this zone will be either partially or 'fully' saturated, even on slopes. Water may descend further to the water-table (where water can seep into the base of a borehole) and then flows as groundwater through an aquifer towards discharge points at lower elevations.

  Excavation and underground works generally suffer from water inflow and dewatering may lead to settlement as a result of the consolidation of fine soils or internal erosion and the loss of fine particles from coarse soils. Such works need careful monitoring and management of water flows.

  Seasonal variations will lead to heave and the settlement of superficial clays and shallow foundations depending on the conditions before construction. Vegetation is the source of transpiration and its presence causes an increase in soil suction and a decrease in water content. The development of tree-root systems (which need moisture) will lead to shrinkage in clays and ground settlement as the tree grows. The clearance of vegetation leads to swelling of clays and heave of foundations. Seasonal variations may also lead to complex groundwater conditions and there may be different water pressures in permeable strata separated by less permeable strata.

- Partially saturated soils, which locally occur to great depths, may be subject to desiccation over a long period of time. Plastic clays tend to heave on wetting when evaporation is prevented by sealing of the ground surface (e.g. by a new building or road). Soil suction can facilitate the construction of temporary steep slopes, excavation and shafts. However, the effects are short-lived and the long-term protection of slopes from erosion is generally required.

- When the annual evapotranspiration exceeds infiltration, water-tables are generally low and are controlled by local stream levels and the presence of permeable strata. In hot, dry areas without tree cover, the upward movement of moisture and evaporation from the ground surface may produce a chemically active zone at ground level. This may lead to situations with aggressive evaporation conditions, as in coastal sabkha or where duricrusts form on dryland surfaces (see Figures 2.2, 2.5 and 3.6).

- Artesian groundwater pressure is driven by high water-tables in adjacent high ground. It is common below the floor of valleys in folded, bedded strata and in glaciofluvial deposits beneath glacial till on the lower slopes of some valleys. Lenticular aquifers may allow perched water-tables to form on valley sides.

Ground profile with a thin, brown, organic-rich soil developed over an alluvial soil of transported sediment, which lies on the rockhead surface beneath which there is little weathering of the dipping rocks.

[right] Fersiallitic, smectite-rich residual soil, with a pale kaolin-rich horizon, developed on young volcanic ash in a Mediterranean climate.

Soil developed in a temperate environment, by the complete breakdown of the underlying shale, aided by plant roots that open fractures.

Profile exposed in a road cutting through dipping beds of pale sandstones and darker shale; a valley in the hillside above has been formed along the outcrop of the weaker shale, but the shale outcrop is hidden, beyond the first bush, by blocks of sandstone fallen from the adjacent outcrops of the stronger rock.

Terra rosa, a red soil of insoluble residues left after solution of limestone in the wet tropics, filling solution-enlarged fissures in the bedrock.

[left] A ferruginous, smectite-rich, black earth developed on young pyroclastic rocks in a wet tropical environment.

[above] Deep contraction fissures formed by desiccation of a clay-rich mud, with blocks about half a metre across.

[left] Variation in peat soils in temperate environments: above; a thin peat soil beneath a wet grassland;
[middle] thick hill peat eroded into deep gullies;
[below] lowland or fen peat being extracted today for fuel, with cut blocks of saturated peat thrown up onto the bank to drain before being taken to storage, where a whole summer is needed to dry them ready for the fire.

A sabkha of clay and silt sediments intergrown with gypsum along a desert coastline flooded at high tides.

A zeugen (or mushroom rock) eroded by wind-driven sand-blasting near ground level in a desert, with salt crystallisation and physical weathering in a lower bed.

Artesian groundwater rising through a borehole due to natural pressure in an underlying confined aquifer.

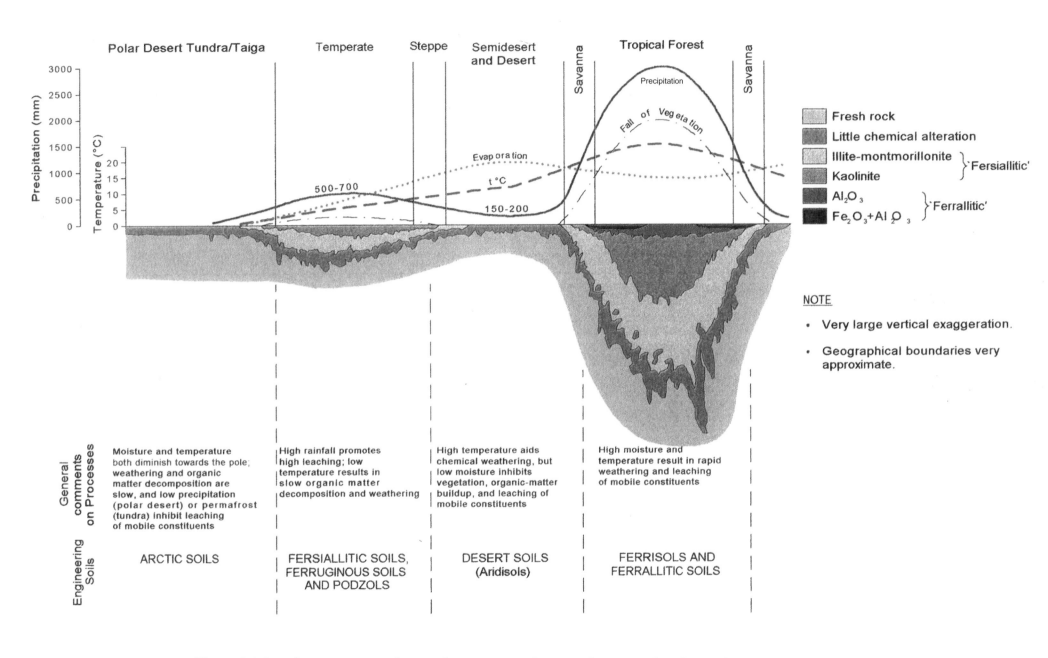

*Figure 2.2 Simple cross-section from pole to equator showing climate and rock weathering characteristics.*

## Climate and rock weathering characteristics from the poles to the equator (Figure 2.2)

The climate, both now and in the past, has a strong influence on engineering design and construction, not only through the production of weathered residual soils, but also on the geomorphological processes that the engineer may expect to encounter on and around a site. The overview of climate, weathering and morphoclimatic regions around the world provided by Figure 2.2 shows the general nature of the influences of the current climates that an engineer may encounter in any part of the world (see Figures 1.1 and 1.2 for descriptions and maps of the major global morphoclimatic zones).

Figure 2.2 is based on an original highly idealized figure in Strakhov (1967). It shows simplified climatic zones from the poles to the equator and, with an exaggerated vertical scale, the weathering profiles of highly weathered soils from the ground surface down to fresh bedrock. The engineering details of the six weathering zones (I–VI) comprising this profile were introduced in Figure 2.1 and are shown in detail in Figures 2.3 and 2.4. For further discussion, see Fookes *et al.* (2005) and Bibliography, Group B books.

The key to the depth of weathering is indicated by the curves of the average annual temperature, the annual average precipitation (snow and rain), an approximation of the annual net evaporation and the length of time for which these conditions have existed. These all show considerable variation between the extremes at the poles and the equator. The relationship between temperature, precipitation and evaporation is the key to understanding Figure 2.2. For example, the tropical forest zone has a high rainfall, warm temperatures (similar daytime and night-time temperatures, i.e. an equatorial climate) and evapotranspiration that is significantly less than the rainfall. In other words, the ground remains warm and moist with only a small evaporative pull, which is conducive to deep chemical weathering of the bedrock. A thick residual soil profile is therefore produced above fresh rock. Fresh rock is only encountered tens of metres below the ground surface unless the residual soils are eroded away.

Contrast this with the semi-desert and desert zones. The average annual temperature appears to be lower than that in the rain forest zone, but the daytime temperatures are typically much higher and the night-time temperatures are often much cooler. Rainfall is low and the net evaporation is much higher than the rainfall. This keeps the ground relatively dry and the small amount of water that occasionally penetrates the ground from storms is quickly lost again to the atmosphere as moisture. Such climatic conditions lead to very slow overall weathering, with almost no chemical weathering (i.e. very little development of residual soils), but there is more mechanical weathering as a result of the large, rapid daily temperature changes that physically split the rocks.

The common processes in the different climatic zones are given below each soil profile in Figure 2.2, which also gives a simplification of the engineering soil terms.

Figure 2.2 requires more explanation of the soil terms used in the key than can be given here. They are directly related to the minerals produced within the soil profile by chemical weathering and are described in specialist books on the subject (e.g. Fookes, 1997b). The value of the figure is that it shows which type of residual soil can reasonably be expected in the different climatic zones. There are many variations depending on the rock type, the details of the climate, the local geomorphological systems and the site-specific patterns of erosion. A comprehensive terrain model as part of a thorough ground investigation of the project site would evaluate all of these factors.

For more on the 'engineering soils' shown in Figure 2.2, see Figures 2.3, 3.1, 3.2, 3.3 and 3.5 (arctic soils); Figures 2.3, 2.4, 2.5, 3.7 and 3.8 (residual soils, with increasing intensity of weathering from fersiallitic, ferruginous, ferrisols – a transitional stage – to ferrallitic); and Figures 3.6 and 3.7 (desert soils).

An engineering geologist may have an input into land use studies for environmental and social impact assessments and will need to be familiar with some of the pedological terms used in agriculture in order to be able to communicate effectively with specialists in other disciplines (International Union of Soil Sciences Working Group, 2006). For example, the residual soils encountered on a project may include one or more of the following FAO–UNESCO terms (Fookes, 1997b):

- *ferralsols*, deeply weathered red or yellow soils with low activity clays (these are equivalent to ferrallitic soils with gibbsite or kaolinite);
- *nitosols*, deep, well-drained red or red–brown soils containing kaolinite or (meta) halloysite clays (these are broadly equivalent to fersiallitic soils);
- *plinthosols*, composed of ferruginous plinthite, petroplinthite or pisoliths (these include groundwater laterites and ferricretes);
- *andosols*, dark-coloured soils containing allophane clay, which typically develop on volcanic ash;
- *vertisols*, heavy clay soils with a high proportion of smectite clays subject to swelling and shrinkage (e.g. black cotton soils).

Other soil types that may occur include *fluvisols* on alluvial deposits, *gleysols* in wetlands, *gypsols* in arid regions and very shallow stony *leptosols* in mountains.

The major soil groups can be further subdivided using various prefixes and suffixes to qualify these terms.

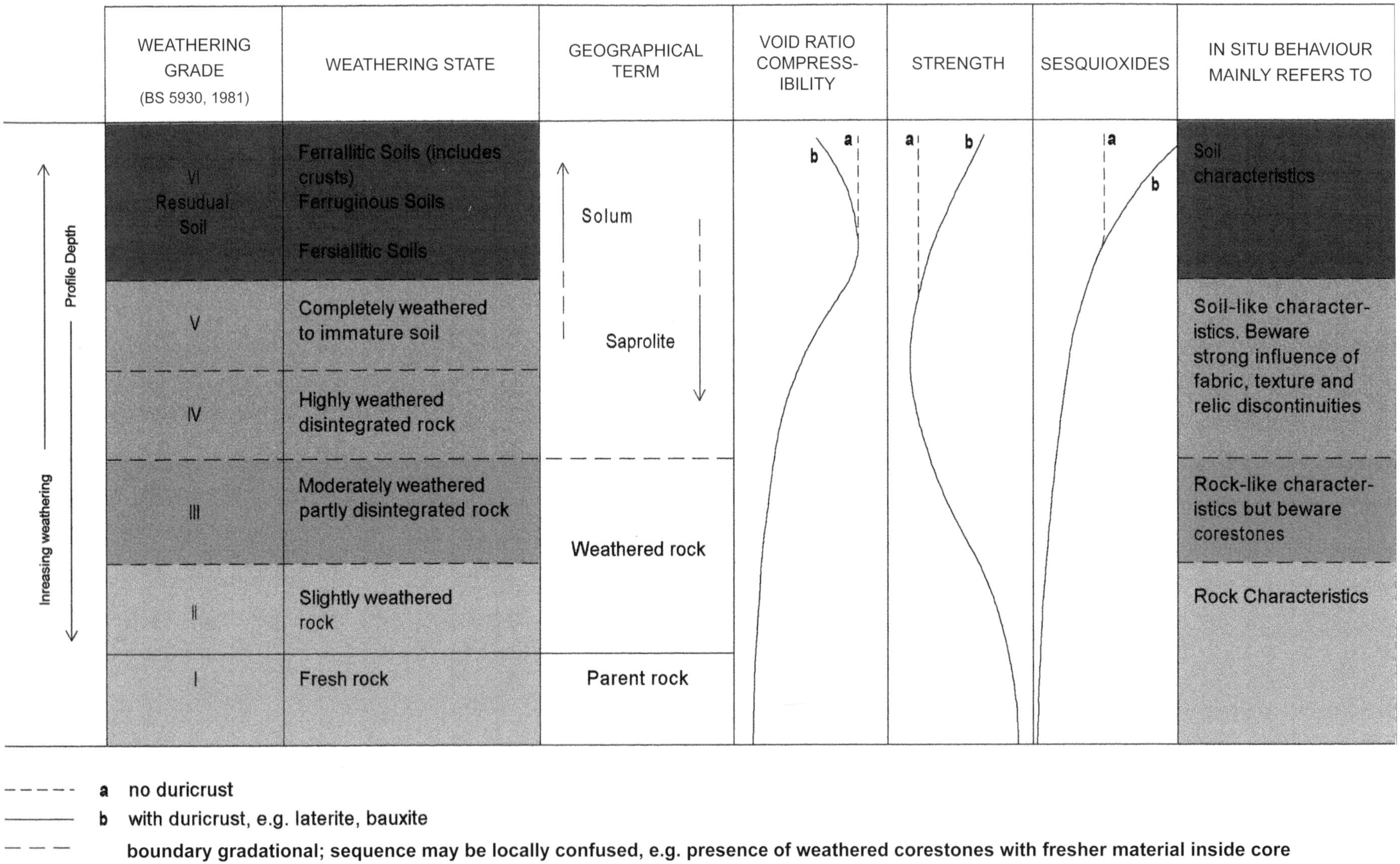

| WEATHERING GRADE (BS 5930, 1981) | WEATHERING STATE | GEOGRAPHICAL TERM | VOID RATIO COMPRESS-IBILITY | STRENGTH | SESQUIOXIDES | IN SITU BEHAVIOUR MAINLY REFERS TO |
|---|---|---|---|---|---|---|
| VI Resudual Soil | Ferrallitic Soils (includes crusts) Ferruginous Soils Fersiallitic Soils | Solum | | | | Soil characteristics |
| V | Completely weathered to immature soil | Saprolite | | | | Soil-like character-istics. Beware strong influence of fabric, texture and relic discontinuities |
| IV | Highly weathered disintegrated rock | | | | | |
| III | Moderately weathered partly disintegrated rock | Weathered rock | | | | Rock-like character-istics but beware corestones |
| II | Slightly weathered rock | | | | | Rock Characteristics |
| I | Fresh rock | Parent rock | | | | |

Profile Depth

Increasing weathering

- - - - -  **a**  no duricrust

———  **b**  with duricrust, e.g. laterite, bauxite

- - -  boundary gradational; sequence may be locally confused, e.g. presence of weathered corestones with fresher material inside core

*Figure 2.3 Conceptual geological and engineering changes in a weathering profile.*

## Conceptual geological changes within a weathering profile (Figure 2.3)

It must be emphasized that Figure 2.3 is a highly idealized compilation of many weathering classifications given in various codes and standards (see Bibliography, Group B books; Figures 2.4 and 2.5). Figure 2.3 is shown here primarily to illustrate the classical sequence of weathering grades I–VI. These are assigned to different stages of weathering, from fresh rock (I) to fully developed residual soils (VI), together with the descriptive terms and in situ engineering behaviour commonly ascribed to each of the weathering grades. The important concept of weathering profiles above fresh bedrock has already been noted in the Introduction and in Figures 1.1, 1.2, 2.1 and 2.2. See also text box on the weathering grade classification.

The identification and characterization of weathering conditions are described in detail in Norbury (2010), who also gives a short history of the approaches to the classification of weathering and background geological and engineering information, amply illustrated by photographs, tables and references. For greater discussion of the various weathering processes, see Bibliography, Group A books; the engineering aspects of weathering are discussed in the books in the Bibliography, Group B, especially Bell (2000).

The land surface is continually modified by weathering and erosion, with a net general lowering over short to long periods of geological time. This is called denudation. All rocks exposed to the atmosphere undergo weathering and weathering processes are commonly divided into three main types.

- *Chemical weathering* brings about decomposition by the breakdown of minerals and the formation of new compounds, with some losses by solution. The main agency for this is water (both rainfall and groundwater) and, in general, chemical processes accelerate at higher temperatures. The typical end-product of chemical weathering is clay. The processes involved in weathering include solution, oxidation, reduction, hydration (the absorption of water molecules into the mineral structure), hydrolysis (the formation of an acid and a base from a salt by the action of water), leaching and cation exchange.

The increase in iron and aluminium oxides and hydroxides as the weathered soil develops is an indication of its weathering maturity, as shown in Figure 2.3 by the sesquioxide curve ($Fe_2O_3$ and $Al_2O_3$).

The precise type (or types) of clay mineral formed by long-term weathering depends on the original rock, the location, the climatic regime, biological influences and the length of time over which the weathering has taken place. Weathering happens slowly all the time, but it typically takes many tens to hundreds of thousands of years to reach a mature state with a thick residual soil profile in equilibrium with the climate. Weathering tends to follow joints and other fractures within the rock, developing from the rock surface inwards. The net result of this at any stage of weathering is a highly irregular weathering front (Figure 2.4). Be aware of any boreholes indicating otherwise.

Not all rocks produce thick weathering profiles. The climatic regime strongly influences the form, rate and depth of weathering. Warm wet climates favour chemical weathering (Figure 2.2). Some rocks, in particular limestones, weather by dissolution in wet climates, with the formation of a karst landscape characterized by underground drainage, caves, sinkholes, dry valleys, thin soils and bare rock outcrops.

- *Mechanical (physical) weathering* is the disintegration of rocks into smaller particles, particularly by the action of large daily and seasonal temperature changes in areas of low rainfall, and by the action of frost and ice. The processes of mechanical weathering include unloading that creates stress relief, allowing fractures to open as the confining stresses of burial are removed by denudation. Also important are thermal changes, wetting/drying and crystallization pressure caused by the growth of new minerals in the pores and fissures of a host rock. In addition, depending on the rock type, the impact of raindrops and abrasion from particles carried in the wind can wear away the surface of rocks.

Cold dry and hot dry climates favour mechanical weathering (Figure 2.2). The shape of landforms in hot deserts tends to be dominated by sandblasting, the type of rock and its susceptibility to sandblasting. In cold climates – such as those towards the poles and very high lands not covered by permanent ice, or where the ground is not permanently frozen – freeze–thaw processes tend to deform the ground surface (Figures 3.1, 3.2 and 5.9). Repeated cyclic freezing breaks off rock fragments, which then litter the landscape.

- *Biological weathering* refers to those mechanical and chemical changes that are directly associated with the activities of animals and plants. When present, microbial activity changes the chemistry of the ground. Burrowing animals may weaken the ground and the growth of tree roots can prise apart blocks of rock, whereas a cover of low-growing vegetation protects the ground from erosion. Vegetation increases the acidity of circulating rainwater because organic products derived from plants are broken down by the action of bacteria and fungi. In general, individual biotic weathering effects are small, but, added together, they can have a significant effect on all aspects of weathering. Biological weathering tends to be more important in warm and wet climates with abundant vegetation.

## Weathering grade classification and its engineering application (after Hearn, 2011)

*Weathering Grade VI*

Residual soil: all rock material converted to soil; mass structure and material fabric destroyed; behaves as a soil; soil mechanics principles to be applied to excavation design; rippable during excavation; when excavated described as fine material; depending on soil characteristics excavated materials treated as common fill (if suitable), treated fill (where removal, mixing or blending is required to allow usage as fill) or unsuitable (cannot be used as fill due to susceptibility to erosion (unless protection is provided), too high a clay content or too low a plasticity; moisture control required during placement. High plasticity clay soils are not uncommon in tropical residual soil profiles and these will have low friction and may be subject to long-term softening as a result of loss of effective cohesion. Failures on 10° slopes in highly plastic residual clays have been recorded.

*Weathering Grade V*

Rock is completely weathered: all rock material is decomposed and/or disintegrated to soil; original mass structure still largely intact, failure may occur on joints; considerably weakened compared to weathering grade IV material; slakes when wet; weathering products and relict structure control strength and stiffness; soil mechanics principles to be applied to excavation design, with a kinematic check required due to the relict structure (e.g. persistent at an unfavourable attitude), rippable during excavation; when excavated described as fine material; depending on soil characteristics excavated materials treated as common fill (if suitable), treated fill (where removal, mixing or blending is required to allow usage as fill) or unsuitable (cannot be used as fill due to susceptibility to erosion (unless protection is provided), too high a clay content or too low a plasticity); moisture control required during placement; potential for loss of structural strength during excavation, haulage, placement and compaction, and potential for loss of strength on wetting. High plasticity clay soils are not uncommon in tropical residual soil profiles and these will have low friction and may be subject to long-term softening as a result of loss of effective cohesion.

*Weathering Grade IV*

Rock is highly weathered: *in situ* rock fabric or texture contributes to mass strength; matrix or weathering products control stiffness; more than 50% of the material is decomposed or disintegrated to soil; remainder forms clasts that cannot be broken by hand and do not readily disaggregate or slake when a dry sample is immersed in water, but which may break down/ degrade over time and are present as a discontinuous framework or corestones 'floating' in a soil matrix; combination of soil mechanics and rock mechanics principles to be applied to excavation and foundation design; typically rippable during excavation but potentially problematic due to presence of boulders/corestones within the soil matrix (blasting of large remnant blocks may be required to break them down to a size that can be excavated and transported); when excavated, described as fine material with some (5–20%) or many (20–50%) boulders or cobbles; may not be suitable as fill due to gap grading e.g., boulders in a fine matrix. A 'mixed fill' category might be required.

*Weathering Grade III*

Rock is moderately weathered: in situ rock framework controls mass strength and stiffness; in excess of 50% of the material forms clasts that cannot be broken by hand but which may break down/degrade over time; shear strength along joints is typically markedly lower than for slightly weathered rock; combination of rock mechanics and soil mechanics principles to be applied to excavation design; potential for kinematic failure may exist; combination of ripping and blasting required for excavation depending on percentage of materials weathered to soils and joint pattern; when excavated described as boulders or cobbles, with some (5–20%) or much (20–50%) fines; behaves as a poor rockfill which requires careful screening of fines and moisture control during placement and compaction. Requires intensive investigation to get a clear picture of sub-surface conditions. Boundaries with adjacent grades irregular.

*Weathering Grade II*

Rock is slightly weathered: there has been some loss of material strength; >90% of materials remain as competent rock; <10% of materials have soil properties; more weathered, weaker materials are located along joints; joint shear strength is typically markedly lower than for joints in fresh rock; rock mechanics principles should be applied to excavation design; potential for kinematic (joint-controlled) failure may exist; blasting required for excavation (depending upon rock type and structure); excavated materials behave as clean, competent, essentially free-draining rockfill (depending upon rock type and structure).

*Weathering Grade I*

Rock is fresh with no visible signs of rock material weathering.

Norbury (2010) proposed five approaches to the description of weathering. His first approach forms the basis of any classification and suggests that the information to be recorded for engineering should include:

- the degree and extent of colour change

- the original strength of the soil or rock and any changes in that strength associated with weathering

- the fracture state of the weathering profile using normally defined terms and measurements, with a specific note of where the fractures are thought to be due to weathering

- the presence and nature of any weathering products described using appropriate soil or rock descriptive terms and a quantification of their extent.

His second and third approaches are for the classification of homogenous medium–strong and stronger rocks and are based on the progressive weathering of intact material. His fourth approach is for the classification of heterogeneous weak rocks (e.g. the Permo-Triassic Mercia Mudstone and the Jurassic Oxford Clay in the UK); his fifth approach is for special cases (e.g. 'chalk').

A massive gabbro slowly weathering by chemical changes that advance inwards from joint faces reached by air and water.

Chemical weathering of a dolerite with production of weak clay minerals first along the open joints 0.5m apart.

Chemical weathering of limestone as the entire rock is removed in solution by water in rills 30 mm wide.

Mechanical weathering of limestone in temperate terrain where frost shattering produces a veneer of broken rock.

Large-scale exfoliation dome in granite with concentric joints created by stress relief; note climbers on the ladder on the left.

Stress-relief fractures parallel to a cliff face.

[left] Granite boulder split by repeated thermal expansion and contraction in a hot semi-desert.

Organic weathering, with a tree root forcing open a joint in limestone.

Mechanical weathering with an outer shell of weathered rock breaking away, left on a granite cliff face, and right on a sandstone tombstone.

A sequence of weathering zones in basalt lavas.

Zone V: completely weathered to a residual soil.

Zone IV: highly weathered, with corestones in a soil.

Zone II: slightly weathered, fractured bedrock.

Zone III is absent as its place is taken by a layer of fine-grained tephra that has weathered more than the lavas.

Organic soil of Zone VI is just visible at the top left.

The profile does not reach down to fresh rock of Zone I.

A shallow landslide that is just one consequence of long-term weathering of a hillside creating a soil that becomes progressively thicker and weaker until it fails, typically when saturated by a heavy rainfall event

Weathered sandstone exposed on a building site; the upper three metres have been removed so that the concrete column bases can be cast in less-weathered, stronger rock beneath.

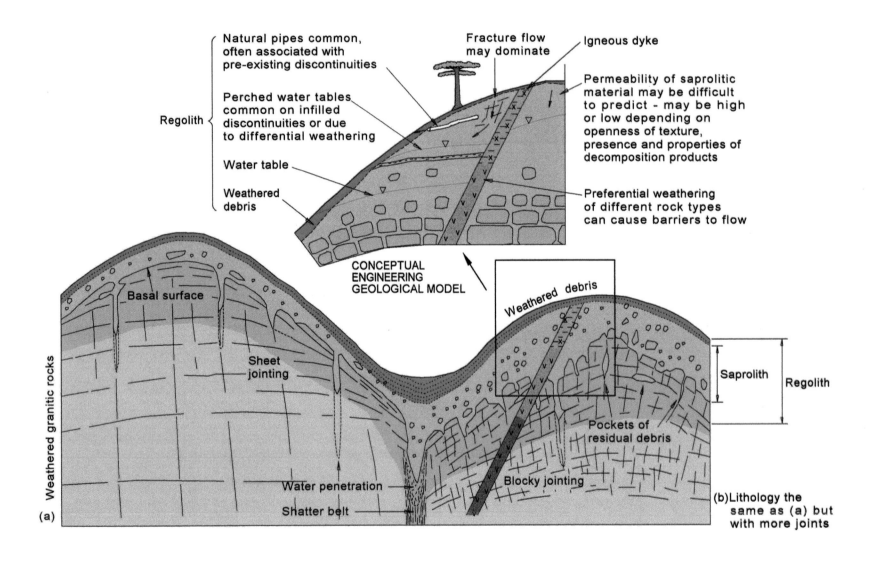

Figure 2.4 Models of tropical weathering: granitic weathering.

## Residual soils: an example of granitic rocks (Figure 2.4)

### MODEL OF DEVELOPMENT OF RESIDUAL SOILS

Figure 2.4 shows the idealized development of tropical residual soils on two types of granite (see also Figure 3.8). The model on the left of the diagram has minimal jointing down which weathering can penetrate. The right-hand model has a greater density of joints, which allows weathering to progress much further under the same weathering conditions, so the saprolith and regolith are thicker. Some idealized water flow situations in the regolith are also shown.

In situ residual soils have a wide range of clay mineralogy, grading and plasticity, with consequent variations in their engineering properties. These properties all depend on the weathering processes and the stage of development of the weathering (see Figure 2.2). The grade of weathering is usually dominated by the amount of unweathered quartz particles present in silicate-rich (acidic) rocks, but this is not true in basic (silicate-poor) weathered rocks. Kaolinitic clays are the typical mineral of residual soils on granitic rocks; iron-rich clays are typical of more basic rocks. Re-cementation of soils may occur (see Figure 2.3).

The soil fabric, texture and the presence of relict discontinuities may dominate the behaviour of undisturbed material in saprolites of weathering grades IV and V. Corestones are usually well developed in granites of weathering grade III and in other rocks with wide joint spacings. Table 2.4.1 is a useful guide for what to record on natural and man-made exposures. Weathered material will always be complicated and highly variable making the ground model challenging.

### RESIDUAL SOILS IN GENERAL

The type and content of clay minerals in residual soils (weathering grades V and VI) progress from (i) fersiallitic (dominated by 2:1 clays) to (ii) ferruginous (kaolinite and 2:1 clays) to ferrisol (a transitional stage, but not always present) to (iii) ferrallitic (kaolinite and gibbsite) groups, depending on the maturity of the weathering process and the climatic condi-

tions. In subtropical climates with a marked dry season, stage (i) is rarely exceeded; in a dry tropical climate development stops at stage (ii) and it is only in humid equatorial climates that stage (iii) is reached (Duchaufour, 1982). The hotter the location, the more the weathering progresses towards the end-stage (Figure 2.2). Broadly, the more mature the clay, the less silica and base elements it contains as a result of their removal by leaching during the weathering process. Table 2.4.2 gives guidance for field recording.

- Engineers commonly describe vertisols (swell/shrink clays) as 'black cotton' soils. They are formed under conditions of poor drainage in warmer climates, have a high plasticity and are typically expansive (usually due to the smectitic clays), with large volume changes on wetting and drying.

- Widespread red-coloured soils are usually formed under good drainage conditions and are very variable, although they commonly have a pronounced structure as a result of the weathering process. Their clay mineral assemblage usually includes kaolinite and they are dominated by the ferruginous, ferrisol and ferrallitic residual soil groups. Their engineering properties depend more on their structure than on the grading and mineralogy and they usually behave as if they are bonded, yielding at a certain stress level. The strain and compressibility depend on this yield as much as on the density. It is difficult to obtain undisturbed samples without destroying the structure. Importantly, the mineralogy and engineering properties may be changed by drying and may even develop a cementation.

These soils are the typical products of weathering in tropical environments and are characterized by the development of iron and aluminium oxides and hydroxides. These compounds, especially those containing iron, are responsible for the red, brown and yellow colours of the soils, each colour reflecting the state of oxidation. They may be fine-grained and may contain nodules or concretions that develop in the matrix when there are high concentrations of oxides. The highest concentrations of

oxides give rise to laterites, a residual ferruginous material generally occurring as a hardened ferruginous crust, which can make a good construction aggregate. The ratio of silica to the sesquioxides of iron and aluminium tends to control their characteristics. A ratio of <1.33 is indicative of laterite soils, a ratio between 1.33 and 2.0 is indicative of lateritic soils, and a ratio >2.0 indicates a non-laterite soil. Such minerals tend to strongly influence the engineering behaviour (see Figure 2.5 and Fookes, 1997b).

*Relict discontinuities* from parent rocks can create planes of low drained strength in the less mature residual soils of grades IV and V. They can also cause slope instability if their extent and inclination are critical. They can be difficult to characterize during site investigation.

*Corestones* of rock with little weathering (commonly within grade III) may cause problems in drilling, piling and excavation; one of the more difficult problems is that they can be misinterpreted as the top of the bedrock.

*Porous soils* with high void ratios are common in dry conditions and may collapse on wetting. Such soils have a high degree of saturation with a low undrained strength and a structure that is easily destroyed. Such soils are likely to form from volcanic rocks (typically fersiallitic andosol clays) and can cause problems for operating plant during excavation and as filling materials because they may liquefy when disturbed.

*Collapsing soils* can occur in residual soils (see Figure 2.3). These soils decrease in volume by rearrangements of their particles with the addition of water with no change in load; this is a result of the destruction of their weak cemented structure. The collapse potential can be evaluated by a flooding consolidation (oedometer) test.

It should be noted that, in general, the red soils that contain clay mineral groups with variable amounts of iron and aluminium sesquioxides can significantly affect the results of laboratory tests, either when dried in situ or in laboratory testing (Fookes, 1997b; Head, 2006, 2011, 2014; Walker, 2012). This can occur as a result of the following processes.

Table 2.4.1 Summary of information to be recorded on exposures (after Norbury, 2010).

| Location/ type of exposure | Information required | Procedure/ Actions |
|---|---|---|
| General | • Important to use a logging proforma because it contains the prompts so that all relevant information is recorded | |
| Trial pits, trenches, natural exposures | • Dates of excavation and logging<br>• Description of each stratum<br>• Depth of stratum changes<br>• Log as many faces as needed to record variability<br>• Sketch geology in each face unless they are all the same<br>• Stability of faces<br>• Description and location of all discontinuities<br>• Levels of water inflow, estimate of rate of flow, rest water levels if achieved<br>• Details of any pumping carried out<br>• Equipment in use including excavator<br>• Subjective ease of excavation<br>• Weather<br>• Plan dimensions and orientation<br>• Location of pit or survey marker<br>• Orientation of long axis of pit or trench<br>• Whether logged in situ or on arisings in excavator bucket<br>• Identify continuity around all four faces<br>• Depth and position of all samples taken<br>• Depth and position of all tests carried out | • Faces are logged by examination of the exposed faces, usually from the surface<br>• Materials are logged on arisings in the excavator bucket<br>• Logging in situ requires the excavation to be supported before the faces can be cleaned of smeared and disturbed materials<br>• Samples are taken from selected strata (ensure samples large enough for proposed testing)<br>• Tests are carried out at required depths or in selected materials, either in situ or on arisings in the excavator bucket<br>• Log faces or arisings, not the recovered samples<br>• Photograph pit in two directions<br>• Include visible scale in all photographs (hammer, ranging rod or survey staff)<br>• Photograph spoil heap(s) |
| Foundation inspections | • Measure and record size, depth and arrangement of foundations<br>• Record location in plan and section related to a grid or permanent markers<br>• Type of foundation construction, eg brick, concrete, wood<br>• The presence and nature of any blinding layers or sealing materials<br>• The underlying and surrounding materials should be described<br>• Describe structural features such as expansion joints, weep holes, ties, damp proof course<br>• Describe any defects or areas of poor quality construction<br>• Describe cracks and joints individually showing aperture and variation in crack width along the length of the crack, infill, surface alteration/ staining.<br>• Note changes in brickwork colour which may indicate a change from engineering bricks to standard type bricks | • A plan view is essential<br>• The location should be referenced to permanent features or site grid<br>• Sections should be drawn to show the various elements clearly<br>• Dimensions should be included on plans and sections to provide legibility and accuracy<br>• Photograph the features revealed by the excavation<br>• Photograph the location of the excavation |
| Large excavated or natural exposures | • Record disposition of different materials<br>• Record the material and mass characteristics of all strata<br>• Record any evidence of natural processes affecting the ground or the site | • Create some form of grid to break face into regular units that can be mapped |

Table 2.4.2 Descriptive scheme for residual soil materials, after Fookes (1997b).

| Parameter | Classification | Procedure |
|---|---|---|
| Moisture | State | Dry = light colour, loose brittle |
| | | Moist = range of colours, neither wet nor dry |
| | | Wet = visible water films |
| Colour | | Use colour charts and describe as for other soils or rocks |
| Strength | Fine soil | Describe consistency |
| | Coarse soil | Describe density |
| | Rock | Describe rock strength |
| Fabric | Origin | Orthic = formed in situ by soil forming processes, e.g. coatings, nodules, peds |
| | | Inherited = relicts of parent material, e.g. lithorelics |
| | Voids | Low, medium and high terms based on laboratory measurements of porosity |
| | Orientation | Strong = most particles are sub parallel |
| | | Moderate = many particles are sub parallel |
| | | Weak = some particles are sub parallel |
| | | None = no particles are sub parallel |
| | Distribution | Porphyritic = matrix is dense |
| | | Agglomeritic = matrix loose or incomplete |
| | | Intertextic = grains embedded in porous matrix |
| | | Granular = no groundmass |
| | Fissuring | Orientation, spacing and character as the standard guidance |
| Texture | | Grain size as the standard guidance |
| Density | | Low, moderate and high terms based on laboratory measurements of density or relative density |
| Apparent behaviour | Remoulded strength | As density above |
| | Durability | Slakes, breaks, chips terms based on field tests |
| | Plasticity | Plasticity described as the standard guidance |
| Mineralogy | | Hand lens examination, carbonate content test |

- *Aggregation of clay particles.* On drying in the laboratory or naturally under appropriate circumstances, aggregates of silt and fine sand size (i.e. not clay-sized) particles can form that significantly coarsen the grade and reduce the plasticity of the soil. This gives misleading test results compared with the in situ field properties.

- The *disintegration* of such aggregated soils should be carried out with care in the laboratory so that the individual particles are separated without crushing or splitting. Weakly cemented soils may be disintegrated by finger pressure, but more strongly cemented soils may need soaking overnight with or without a dispersant.

- *Irreversible variable changes in plasticity on drying.* This can occur, depending on the amount of drying, in soils that become less plastic because stronger bonds are created between the particles so that they resist penetration by water.

- *Loss of in situ water of hydration on drying.* This occurs on oven-drying at 105°C (the standard oven temperature in laboratories in temperate climates) and changes the character of the materials. Lower oven temperatures are required for residual soils in tropical climates (Walker, 2012).

[across the top] Granites exposed in a semi-desert terrain:
[above left] laterite crust breaking away from fresh granite;
[above] a thick laterite formed over a granite outcrop;
[right] corestones of granite from weathering zone III left behind after the intervening soil has been eroded away.

Spheroidal, "onion-skin" weathering in igneous rocks:
[above] with peeling shells each about 5 cm thick in granite;
[right] with shells each just a few millimetres thick in basalt.

[left] Deeply weathered granite exposed on a construction site in a wet tropical environment; angular blocks of clean rock remain between zones of red soil formed by weathering down fractures; more highly weathered soils more than 10 metres deep have already been removed.

Granite, forming distant jagged frost-shattered peaks and rounded boulders in front (grade III) where soil has been washed away to leave the weathered corestones.

Two nearly spherical corestones of discoloured granite, formed beneath the surface before the weathered soil along the joints was washed away.

A thick, red, residual, kaolinitic soil of weathering grade V, which is the remains of a granite in a tropical environment.

Contrasting styles of weathering in rocks other than granite. [left] Sandstone, which breaks down to a mass of loose sand with small sandstone blocks, and ultimately to a soil of almost pure sand. [middle] Sequences of basalt lavas commonly include horizons of red, fine-grained material a metre or so thick, known as boles; these were soils formed by weathering in the intervals between successive lava flows. [right] A red, iron-rich soil produced by weathering of basaltic lavas with more iron minerals than in a granite.

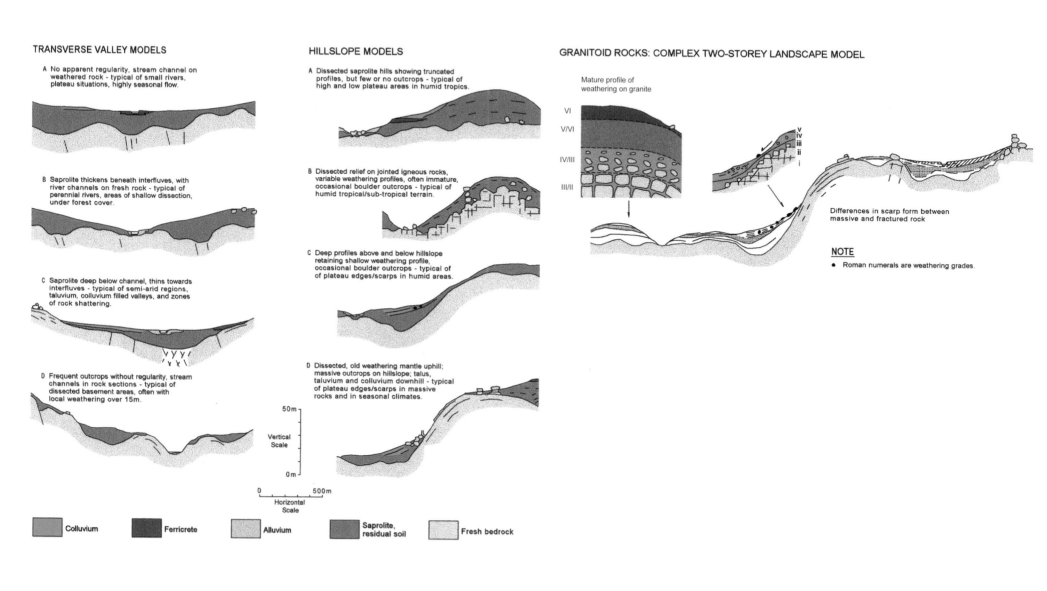

**TRANSVERSE VALLEY MODELS**

A  No apparent regularity, stream channel on weathered rock - typical of small rivers, plateau situations, highly seasonal flow.

B  Saprolite thickens beneath interfluves, with river channels on fresh rock - typical of perennial rivers, areas of shallow dissection, under forest cover.

C  Saprolite deep below channel, thins towards interfluves - typical of semi-arid regions, taluvium, colluvium filled valleys, and zones of rock shattering.

D  Frequent outcrops without regularity, stream channels in rock sections - typical of dissected basement areas, often with local weathering over 15m.

**HILLSLOPE MODELS**

A  Dissected saprolite hills showing truncated profiles, but few or no outcrops - typical of high and low plateau areas in humid tropics.

B  Dissected relief on jointed igneous rocks, variable weathering profiles, often immature, occasional boulder outcrops - typical of humid tropical/sub-tropical terrain.

C  Deep profiles above and below hillslope retaining shallow weathering profile, occasional boulder outcrops - typical of of plateau edges/scarps in humid areas.

D  Dissected, old weathering mantle uphill; massive outcrops on hillslope; talus, taluvium and colluvium downhill - typical of plateau edges/scarps in massive rocks and in seasonal climates.

50 m

Vertical Scale

0 m

0          500 m

Horizontal Scale

**GRANITOID ROCKS: COMPLEX TWO-STOREY LANDSCAPE MODEL**

Mature profile of weathering on granite

VI

V/VI

IV/III

III/II

Differences in scarp form between massive and fractured rock

NOTE

• Roman numerals are weathering grades.

| Colluvium | Ferricrete | Alluvium | Saprolite, residual soil | Fresh bedrock |

*Figure 2.5 Examples of valley and hill-slope models in tropically weathered terrains.*

68

## *Valley, hill-slope and surface models in weathered terrain (Figure 2.5)*

### MODELS OF VALLEYS AND HILL-SLOPES

Figure 2.5 shows a variety of small simplified hill-slope and valley models developed on granites and other strong rocks. The models show how significant differences can occur over a period of geological time in a common set of geomorphological and geological circumstances. Lateral variations in tropical residual soils result from two main factors:

- a spatial pattern of variable depths of weathering, some to many tens of metres, forming a covering mantle
- catenas (repeated sequences of soil profiles developed on slopes) superimposed on this irregular, weathered mantle.

The individual model examples in Figure 2.5 should be sufficient to give an indication of the likely history of the landscape and of its previous and current climate settings. Colluvium (fine hill-slope debris) can also include taluvium (coarse slope debris). Note the difference in the vertical and horizontal scales; there is much vertical exaggeration (Fookes, 1997b).

### LANDFORMS AND CLIMATES

The small landforms are the products of long periods of development, typically up to millions of years. Their development is therefore spread over various climatic regimes and geomorphological influences on a predominantly stable shield (or craton). This is an ancient landscape that is slowly being weathered and eroded down to low-level plains. The climates range from temperate Mediterranean to tropical (see Figures 1.1 and 1.2). Most of the models are found in middle Africa, but also illustrate slope development in other stable, continental landscapes, such as large parts of Australia, India and central southern USA. They are not models for cold or very dry climates.

The models are characteristic of residual landforms and deposits commonly found in modern savannas (see Figure 3.7) and adjacent areas. The savannas are typified by a sharp contrast between more or less bare bedrock outcrops and rocks with mantled surfaces, and between steep hill-slopes and shallow pediments (i.e. long concave upward slopes at low

angles) extending from the base of steeper mountains. Massive granite outcrops typically provide strong, weathering-resistant, upstanding land with a variety of strong, but less resistant, rocks remaining, mantled with debris even on slopes exceeding 35°.

The characteristics of pediments are thus varied: some are thinly mantled and cut across various rock types and others are decomposed by their long history of weathering developed over less resistant rocks. The slope angles of true old pediments seldom exceed 7°, but profiles on overlying debris, not uncommon in present day savannas, may exceed 15°.

Established relationships between climate and weathering products (McFarlane, 1983) may be a poor guide to saprolite properties in deeply weathered mantled areas developed in former, more humid environments. Many mantles are dominated by kaolinite clays (stable, not moisture-sensitive, i.e. engineer-friendly) in the upper parts of the residual soil profiles, with micaceous and interstratified clay minerals increasing with depth. Smectite clays (with a high shrink-and-swell potential on changes in moisture content, i.e. engineer-unfriendly) are rarely developed beneath forests, but increase in favourable locations as the amount of rainfall decreases. Kaolinite saprolites tend to characterise the upper parts of free-draining slopes, but the lower and wetter valley floor are typically smectites.

Seasonal tropical climates experiencing periods of intense prolonged rainfall typically undergo shallow debris slides on steep slopes, and deeper landslides may be generated on flatter slopes in clay-rich saprolites. Debris flows and mudflows are rarely recognized with certainty, but extensive sheets of colluvium on sloping ground between granite hills have been interpreted as mudflows and are important in evaluating engineering projects. Debris fans also occur.

A gradual decline in the water-tables through geological time, commonly as a result of uplift, results in the hardening of residual soils, often with the extensive development of nodules. These become concentrated in the upper parts of vertical profiles as finer and less dense materials are washed downslope. Iron crusts, i.e. ferricretes, are widespread in wetter savannas, where they may be relicts of wetter, Tertiary climates.

The two-storey landscape model at the bottom of Figure 2.5 (of similar appearance to hill-slope model D) displays complex patterns of weathering resulting from uplift and the ancient creation of two separate landforms: the newer uplifted plateau and the older lower landform. The surviving fragments of the older landform have thicker protective surface duricrusts and can often be correlated with continent-wide erosion surfaces.

### DURICRUSTS

The *ferricretes* are members of the duricrust family, which are indurated (hardened) horizons at or near the ground surface. Other members of the family include *alcretes* (or alucretes) which, like ferricretes, are formed by the relative accumulation of iron and/or aluminium oxides in the soil as more mobile compounds are leached out of the weathering profile. They tend to be associated with hot, high rainfall climates. *Silcretes* (silica), *calcretes* (calcium carbonate), *dolocretes* (magnesium carbonate) and weaker *gypcrusts* (gypsum) also form through accumulation in the weathering profile. This accumulation may occur as the result of capillary rise, downward percolation or the throughflow of solute-rich groundwater. These are considerations for geologists in the development of the project site model.

Duricrusts occur widely in today's hot deserts and it is important to evaluate them as the ground investigation develops (Figure 3.6). Desert coasts tend to develop gypcrusts (with sulphates from seawater). Cretes (cf. 'concretes') are generally stronger than the underlying materials (which may be porous due to leaching) and hence tend to armour the landscape by forming a hard cap (or carapace) to flat-topped hills and plateaus. They may be greater than 2 m thick and massive, although joints may be widened by solution; crusts (cf. 'pie crusts') tend to be thinner and weaker and are often transient features. Cretes may have significant cost implications for excavation operations and may be sources of construction stone and aggregate. Thicker, stronger cretes may be ripped by heavy equipment, with a hammer on standby (Pettifer and Fookes, 1994); they may need blasting. Attention must be paid to the possibility of contamination by chloride and sulphate salts that are aggressive to engineered structures.

[above] A coarse talus slope of blocks of strong sandstone that have fallen from the cliff above.

[left] Talus slopes that have grown large enough to coalesce into a wide apron of debris in a temperate environment; bracken cover in the central zone shows that it is now less active than the bare debris slopes on each side.

[below] A bajada is a foot-slope of coalesced and often inactive alluvial fans that is common in semi-arid mountain terrains.

A complex slope on dry high mountains, with rock faces weathering above active talus slopes that continue down into alluvial fans with shallower surface profiles.

A large alluvial fan where a steep mountain valley reaches a lowland; the stream deposits new sediment until it slips sideways repeatedly to newer radial courses.

Two large mudflows within temperate environments: [above] on a steep hillside undercut by weathering and erosion of mudrocks beneath a strong sandstone cap; [below] on a clay slope repeatedly undercut by marine erosion between stable cliffs in stronger sandstone.

Duricrusts formed in semi-arid environments:

[above right] a calcrete that is effectively a case-hardened surface layer on bedded limestone;

[right] a conspicuous calcrete formed by lime-saturated groundwater rising through a gravel (note tyre tracks);

[below] silcrete cementing a surface layer, and increasing the strength of a coarse gravel.

[right] A small bornhardt of granite in a savanna.

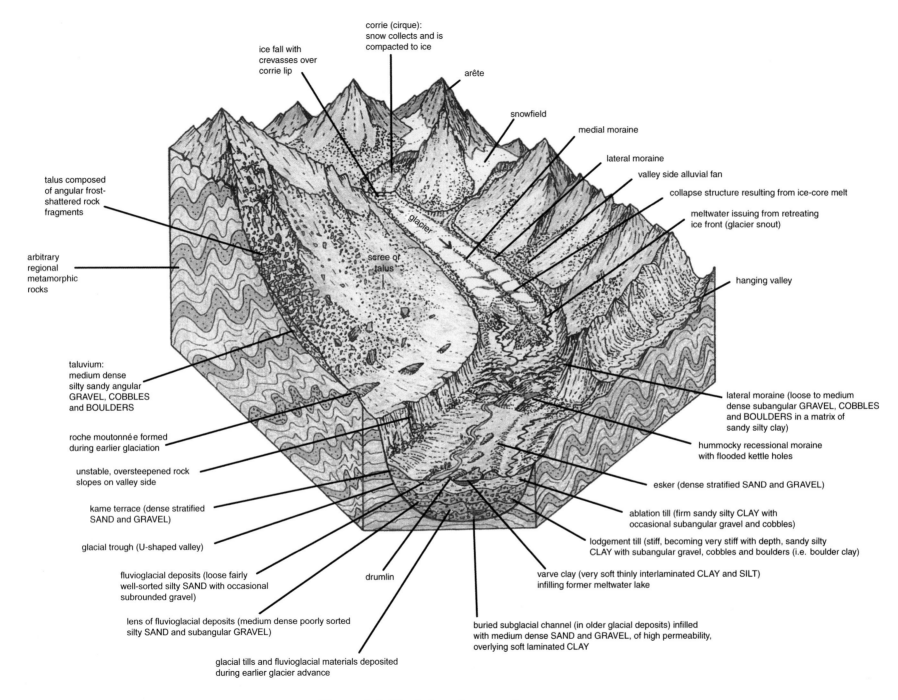

corrie (cirque):
snow collects and is
compacted to ice

ice fall with
crevasses over
corrie lip

arête

snowfield

medial moraine

lateral moraine

valley side alluvial fan

collapse structure resulting from ice-core melt

meltwater issuing from retreating
ice front (glacier snout)

talus composed
of angular frost-
shattered rock
fragments

arbitrary
regional
metamorphic
rocks

glacier

scree or
talus

hanging valley

taluvium:
medium dense
silty sandy angular
GRAVEL, COBBLES
and BOULDERS

lateral moraine (loose to medium
dense subangular GRAVEL, COBBLES
and BOULDERS in a matrix of
sandy silty clay)

roche moutonnée formed
during earlier glaciation

hummocky recessional moraine
with flooded kettle holes

unstable, oversteepened rock
slopes on valley side

esker (dense stratified SAND and GRAVEL)

kame terrace (dense stratified
SAND and GRAVEL)

ablation till (firm sandy silty CLAY with
occasional subangular gravel and cobbles)

glacial trough (U-shaped valley)

lodgement till (stiff, becoming very stiff with depth, sandy silty
CLAY with subangular gravel, cobbles and boulders (i.e. boulder clay)

fluvioglacial deposits (loose fairly
well-sorted silty SAND with occasional
subrounded gravel)

drumlin

varve clay (very soft thinly interlaminated CLAY and SILT)
infilling former meltwater lake

lens of fluvioglacial deposits (medium dense poorly sorted
silty SAND and subangular GRAVEL)

buried subglacial channel (in older glacial deposits) infilled
with medium dense SAND and GRAVEL, of high permeability,
overlying soft laminated CLAY

glacial tills and fluvioglacial materials deposited
during earlier glacier advance

*Figure 3.1 Glacial environments.*

# *Part 3. Basic geological environments influencing engineering*

## *Glacial environments (Figure 3.1)*

Although Figure 3.1 specifically shows a valley glacier, this text refers to glacial landscapes in general. At present, ice caps and glaciers cover about 10% of the Earth's surface and produce some of the most complex environments on the planet (see Introduction, Figures 1.1 and 1.2; also Fookes *et al.*, 2005, and Bibliography, Group A books).

The area covered by glaciers during the Quaternary fluctuated between very extensive glaciations, covering as much as one-third of the Earth's land surface, and limited glaciations similar to present day conditions. Glacial landscapes, both now and in the past, are essentially high-latitude and high-altitude environments. Glaciers affect adjacent (proglacial) non-glaciated landscapes by controlling the nature of stream systems, lakes, coastal environments and wind systems.

In high-latitude regions the glacial environment is commonly bordered by tundra, a fragile mixture of bare ground with varying degrees of lichen, moss and low bush cover, giving way to sparse tundra woodland and then to taiga conifer forest and muskeg (bog soil, extensive in Canada). Proglacial regions are progressively less affected by perennially frozen ground (permafrost) and by the wide range of periglacial processes the further they are from glacial terrains (see Figures 1.1, 1.2, 3.2 and 5.9).

### Glacial landforms

Valley glacier systems are highly variable and may include glaciers mantled with debris, rock glaciers and hanging glaciers. Debris-mantled glaciers produce large lateral and terminal (or recessional) moraines dumped by melting ice (Figure 3.1). These glacial environments are characterized by the presence of supraglacial, subglacial, proglacial, periglacial and paraglacial ice-marginal sediments, which can be identified in the landforms. The engineering characteristics of materials from these environments include variable grading, a low bulk density and a low shear strength, although some

glacial processes (e.g. lodgement tills) may produce materials with a high bulk density and a high shear strength even when the grading is highly variable. For further reading, see Bibliography, Groups A and B books.

The morphological forms of *ice sheets* are highly variable, ranging from small cirque glaciers to very extensive continental ice sheets. Large versions of the latter extended over the high- and mid-latitude regions of North America and Europe during the Quaternary glacial advances, although the ice sheets did not reach the maximum extent in all regions. As a result, characteristic suites of glacial landforms and sediments dominate extensive areas of these regions. Many of the landforms were produced subglacially beneath a thickness of several kilometres of glacial ice. Their internal structures and composition are typically complex and consist of tills (dumped by ice; see Table 3.1.1), glaciofluvial sediments (deposited by melt water streams within or on the surface or sides of the ice) and glaciolacustrine sediments (formed in proglacial lakes that may have covered large areas). Some of the till sheets were moulded by ice action into rounded, elongated mounds known as drumlins.

Many landforms that originated in glacial and proglacial environments are named after their mode of formation, including kames and esker ridges (deposited from water), hummocky sheet moraines (depositions of rock debris scraped from the land by glaciers) and kettle holes left where blocks of ice, detached from a glacier, melted. The advancing glacier front may have overridden or pushed previously deposited sediments to form folds and faults within them. Moraines and other glacial deposits tend to be heterogeneous rubbly materials that include angular blocks of rock, boulders and pebbles, and also rock that has been ground down to clay grade. Water-sorted deposits tend to be better sorted into gravels, sands and silts. The engineering characteristics largely relate to grading and compaction.

- The erosion, deposition and deformation of rocks and sediments occur as a consequence of the continuous movement of glaciers or ice sheets as they advance and retreat. Glacial erosion occurs as a direct consequence of the melting and refreezing of glacial ice on the underlying ground, with abrasion by particles carried on the ice and in the subglacial melt water streams. Glacially eroded features vary in size from millimetre-scale striations and friction cracks to large bedforms such as channels, depressions and roches moutonnées on the 100 m scale.

  The debris produced by glacial erosion may be incorporated into glacial ice sheets at or near the base of the glacier as well as on its surface. Debris entrained in the ice is deposited at the front of the glacier or at its base during melting. Many attempts have been made to classify these till materials and the generally accepted classifications are based on the mode of deposition and include lodgement till (deposited from the base of the ice), melt-out tills (uncompacted debris) and flow tills (deposited from the ice sheet by flow processes).

- Glaciofluvial landforms are formed from debris deposited by water draining through the glacial system, either on or beneath the ice. Glaciofluvial water is usually concentrated in channels eroded into the moving ice, making it difficult to identify the past or present courses of the melt waters and creating fundamental problems in reconstructing the glacial hydrology. Much of this sediment is carried to the proglacial environment, where it is deposited in braided channel systems that may form extensive outwash fans. Large areas of eastern England are covered by such sediments.

- Different types of glacial lakes and overflow channels are commonly associated with glacial environments. Subglacial lakes can occur under ice sheets, notably where the ice overlies active volcanoes, such as in Iceland. Proglacial lakes form primarily behind terminal and lateral moraine ridges and ice-dammed valleys. A complex assemblage of sediments may be deposited in these lakes, including deltas, turbidites and laminated deposits. Such deposits may slump and fault as the ice retreats. Thaw lakes occur in periglacial zones.

- Strong winds are associated with glaciated environments because of the high-pressure systems that develop over these regions. Such winds can reach very high velocities and are capable of moving surface sands and silts well beyond the glacier front, depositing them as metastable sheets of sand and loess (silt). Loess is especially extensive across much of northern continental Europe, North America and western China.

### ENGINEERING IN GLACIAL ENVIRONMENTS

From an engineering point of view, it is essential to distinguish between the modern glaciated terrain of active glaciers and the glaciated landscapes of former glaciations. In glacially active terrains, the main types of glacial hazard are glaciers that are expanding or surging and the outburst floods (known as jökulhlaups) released from glacial lakes. Ice avalanches can break off from glacier snouts, but rarely affect populated areas or infrastructure (see Table 3.1.2).

In formerly glaciated terrains, the bulk properties of the principal glacial and glaciofluvial sediments are very variable on both large and small scales. Furthermore, their properties vary in response to their degree of alteration by processes including weathering and mass movement, the hydrological characteristics of the deposits and local variations between erosional and depositional landforms. Construction projects on glacial deposits require a model-making approach that attempts to understand the complexity of the terrain to facilitate accurate ground investigations on the materials and their forms (Trenter, 1999; see also Part 4).

Table 3.1.1 Some characteristics and properties of different types of till.

| Till type | Formation | Particle size distribution | Mesofabric | Microfabric | Bulk density | Shear strength |
|---|---|---|---|---|---|---|
| Glaciotectonite | Subglacially sheared sediment and bedrock | Poorly sorted | Moderate | Microshears | Moderate | Low |
| Comminution | Subglacially crushed and powdered local bedrock | Poorly sorted skewed towards fine | Moderate | Microshears | Moderate | Moderate |
| Lodgement | Subglacially plastered glacial debris on a rigid or semi-rigid bed | Poorly sorted | Strong up-valley dip | Microshears | High | High |
| Deformation | Subglacially deformed glacial sediment | Poorly sorted | Strong | Microshears | Moderate to high | Moderate to high |
| Meltout | Glacial sediment deposited directly from melting ice | Poorly sorted | May be stratified | Microshears | Low | Low |
| Sublimation | Glacial sediment deposited directly from sublimated ice | Poorly sorted | Preserves ice foliation | Microshears | Low | Low |
| Flow till* | Sediment deposited off the ice by debris flow processes | Poorly sorted | Downslope | Few | Low | Low |

* This is classed as a debris flow rather than a till, but it is included here for comparison.

Table 3.1.2 Types of glacial and glacially related hazards (adapted from Richardson and Reynolds, 2000).

| Category | Hazard event | Description | Time scale |
|---|---|---|---|
| Glacier hazards | Avalanche | Slide or fall of large mass of snow, ice and/or rock | Minutes |
| | Glacier outburst | Catastrophic discharge of water under pressure from a glacier | Hours |
| | Jökulhlaup | Glacier outburst associated with sub-glacial volcanic activity | Hours–days |
| | Glacier surge | Rapid increase in rate of glacier flow | Months–years |
| | Glacier fluctuations | Variations in ice front positions due to climatic change, etc. | Years–decades |
| Glacier hazards as above, plus: | Glacier lake outburst Floods (GLOFs) | Catastrophic outburst from a proglacial lake, typically moraine dammed | Hours |
| | Débâcle | Outburst from a proglacial lake (French) | Hours |
| Related hazards | Lahars | Catastrophic debris flow associated with volcanic activity and snow fields | Hours |
| | Water resource problems | Water supply shortages, particularly during low flow conditions, associated with wasting glaciers and climate change etc. | Decades |

[above] Valley glaciers emerge from a small summit icefield, and descend into larger valleys that were eroded between sharp arête ridges during the Quaternary glaciations.

[above right] The deep U-shaped profile of a valley that carried a powerful glacier during the Quaternary.

[left] Nunatak mountain tops that project through an ice sheet where it is flowing into major outlet glaciers.

[left] Glacial striae scored by ice-dragged boulders into a rock outcrop 5 metres long.

[below] The montane hollow was the site of a small cirque glacier in the Quaternary, when a distant ice sheet fed a glacier along the valley below right.

Medial moraines mark convergence of valley glaciers.

A fiord – a deep glaciated valley now invaded by the post-glacial rise of the sea.

Moraines in a Himalayan valley; the low debris hills in the foreground are terminal moraines a few hundred years old, and the great bank of debris beyond is the lateral moraine from the last major Quaternary glaciation.

Proglacial lake inside a retreat moraine; [right] a succession about 2 metres thick of varved silts from a lake now drained.

Hills formed as glaciofluvial eskers inside a glacier and [right] exposed in a quarry, their sands and gravels draped over till.

Drumlins – formed of till that was remoulded beneath a moving ice sheet and [right] isolated erratics left behind by an ice sheet.

Multiple retreat moraines left by a small glacier lobe that melted away in stages, towards the left, during the global climatic warming that ended the Little Ice Age event a few hundred years ago.

Till – unsorted glacial debris;

[left] on the top of a glacier that is wasting away in the melt zone near its snout;

[right] left by a Quaternary ice sheet as an extensive and thick sheet moraine that lies over unexposed bedrock.

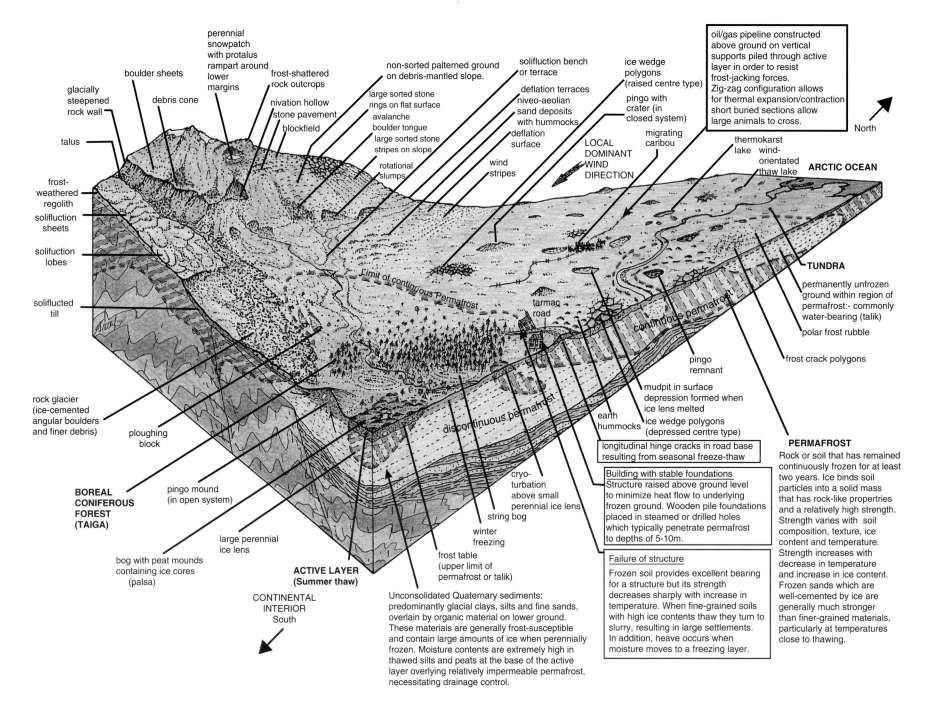

*Figure 3.2 Periglacial terrain model.*

## Periglacial environments (Figure 3.2)

The word periglacial was introduced to provide a term for the cold conditions that bordered Quaternary ice sheets. By general usage it has come to mean a much broader area in which the temperature regime includes alternations between ground freezing and thawing; these environments occur at either high latitudes or high altitudes, or both. About 35% of the Earth's land surface is affected by frost action and about 20% of the present day land area has permafrost beneath its surface.

Permafrost is ground ice and frozen ground beneath a thin active surface layer in which summer thawing and winter freezing occur. The temperature of the frozen ground has to continuously remain below 0°C for more than two years to be called 'permafrost'. These conditions are of great significance to engineering in both active and relict terrains. They lead to many periglacial landforms, including patterned ground, frost mounds and pingos, thermokarst, altiplanation terraces, block fields and rock glaciers. This large and important subject for ground engineering is dealt with further in Figures 3.5 and 5.9.

Permafrost can be divided into four main categories: subsea permafrost, alpine permafrost, continuous permafrost and discontinuous permafrost. Continuous permafrost requires the mean temperature to be below –8°C and discontinuous permafrost requires a temperature between –1°C and –8°C. Continuous and subsea permafrost are currently limited to high latitudes, but alpine permafrost is a feature of high altitudes throughout the world. During the glacial cold periods of the Quaternary, periglacial conditions, including permafrost, extended much further from the polar regions towards the equator than they do today. The former presence of periglacial conditions can be identified by patterned ground and numerous other features left in and on the ground. These relict periglacial features extend over much of the high- and mid-latitude countries (see Figures 3.5, 5.1 and 5.9). Such features can significantly affect ground engineering. For background reading, see Fookes *et al.* (2005) and Bibliography.

### Periglacial processes and landforms

Periglacial landscapes are an amalgam of forms created by processes uniquely related to periglacial conditions, including freezing and thawing, mass wasting and nivation. More normal geomorphological processes driven by wind, water and biological activity are also in evidence in the landscape, but even these forcing agents are usually modified to some extent by the periglacial environment.

The importance of ground freeze–thaw oscillations derive from the unique properties of water and its volume changes as it freezes and thaws. Freezing of the ground, associated with the formation of ground ice, leads to expansion of the soil. Thawing promotes liquefaction and contraction, leading to solifluction and subsidence.

Frost action, a result of cyclic seasonal freezing and thawing, includes frost-wedging (shattering, scaling, splitting), frost-heaving, frost-creeping, frost-sorting, nivation and solifluction. Solifluction describes the down-slope movement of saturated ground in general; gelifluction is the specific term when it is driven by freeze–thaw action. Frost-wedging is commonly the main physical weathering process in cold regions and involves the fracturing of rock as water freezes in its cracks and pores. Frost-heave is the displacement of rock or soil by water as it freezes in the ground. On freezing, there is a 9% expansion in the volume of pore water in the soil, although much heaving is due to the formation of lenses and flakes of ice of varying thickness.

Solifluction (gelifluction) is the predominant form of periglacial mass movement in modern periglacial areas and its products are evident in the more extensive former periglacial areas, where it has left relict features of engineering significance (Hutchinson, 1991). Although not limited to areas with frozen ground, solifluction is widespread on even low slopes because the frozen (and hence impermeable) subsurface ensures a saturated upper layer that therefore remains with excess pore pressure in the thawing soil. This promotes the downhill slope movement of solifluted material. The rates of movement are affected by the degree of the slope, the soil texture, the depth of the thaw, the water content and the extent of vegetation. Movement is generally slow, although occasionally rapid failure can take place through detachment slides of the active layer.

Solifluction was widespread during the Pleistocene and affected all of Britain, so its effects today are relict. Such solifluction material, known as head, can be reactivated with further down-slope flow and is likely to have a lower bearing capacity and higher compressibility than the parent stratum. It also commonly contains shear planes and is therefore prone to modern landslips when undercut.

Other geomorphological processes, driven by wind and running water, are also active in periglacial regions. Water is especially abundant in the summer snowmelt season and typically accumulates between the many hummocks that characterize the surface before concentrating as surface flow and streams, and even as extensive flooding. For example, the Colville River (the largest in northern Alaska) carries about half of its total annual flow during a three- to four-week period. It is during such flooding that much of the fluvial activity in periglacial areas occurs. Thaw lakes can be common (aligned with the wind direction) in the summer melting of the active layer.

Wind is highly variable in different periglacial regions and its effects depend on both the direction and duration of flow as well as the flow velocity. Most of the ground is protected by snow cover during deep winter, but the wind creates drifts and snow ridges; snow crystals become an effective erosive tool. Wind erosion, transport and deposition are common in snow-free periods. River bars, outwash plains and other non-vegetated areas become sources of sediment that is subsequently redistributed by the wind into sand sheets and loess.

Ground ice, a nearly ubiquitous phenomenon in periglacial areas, creates many engineering problems, especially during periods of freeze and thaw, affecting buildings and infrastructure, particularly where permafrost is inadvertently thawed. Except in thaw-stable gravels, conservation of the permafrost therefore has to be the key concern during both construction and the life of the project. Buildings, roads and pipelines are especially sensitive (see Figure 5.9 and Bibliography, Groups A and B books).

[left] The environment of the tundra, with no normal trees.

[above] Treeless tundra across the Arctic Circle.

[above right] The taiga environment with only a scatter of black spruce.

[far left] Outwash plain of glaciofluvial gravels carried by meltwater from a glacier out into the periglacial zone that surrounds the ice cover.

[left] The surface layer of frost-shattered rock fragments, widely known as felsenmeer [meaning stone sea], in a mountain periglacial zone.

[upper right] Frost-shattered rock debris accumulated as scree (or talus).

[right] A rock glacier of frost-shattered debris, with interstitial ice that lets it creep downhill.

[above left] Patterned ground of stone polygons formed by annual re-growth of ice crystals heaving the larger stones away from growth centres as the ground freezes.

[above] Stone stripes that are effectively stone polygons of sorted debris greatly extended by downhill soil creep.

Ice wedges that form large polygons within the active layer are picked out by plant growth on their finer soil.

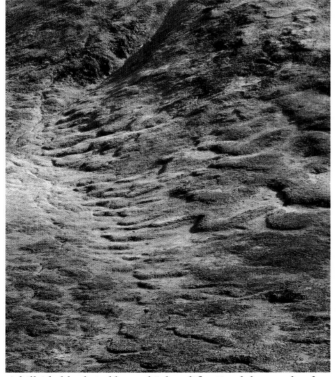

Summer canals over thawed ice wedges, beside a gravel pad that conserves the permafrost beneath the buildings.

A hillside blanketed by multiple solifluction lobes, each a few metres thick, that are developed by slow down-slope creep of annually thawed soils within the active layer.

Timber piles that were placed through the active layer and founded in permanently frozen stable ground; the wooden building on them was then burned down, since when 50 years of solifluction and creep of the hillside soils have rotated the piles above their stable toes.

Ground subsidence over large ice wedges thawed when invaded by warm summer drainage water from a road.

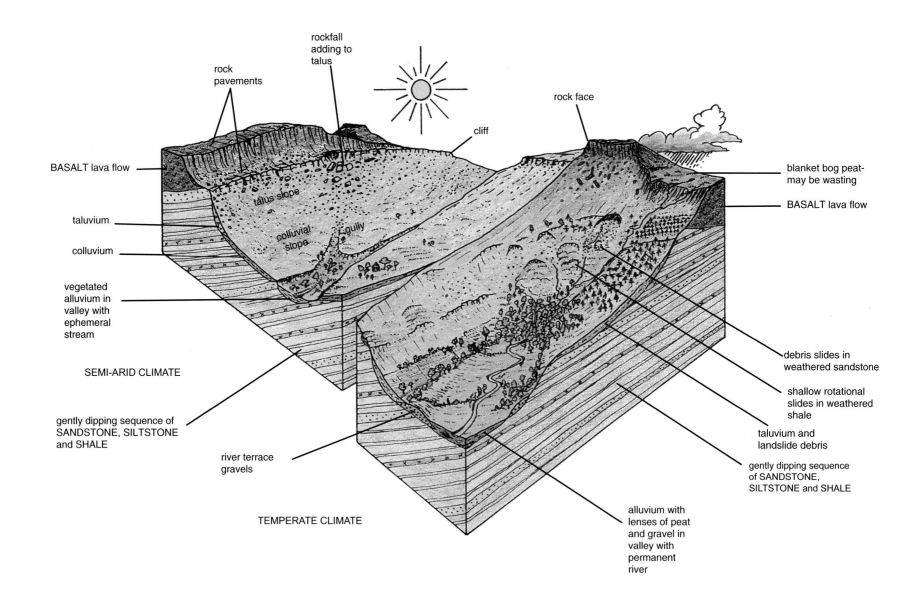

rockfall
adding to
talus

rock
pavements

rock face

cliff

BASALT lava flow

blanket bog peat-
may be wasting

talus slope

BASALT lava flow

taluvium

colluvial
slope

gully

colluvium

vegetated
alluvium in
valley with
ephemeral
stream

debris slides in
weathered sandstone

SEMI-ARID CLIMATE

shallow rotational
slides in weathered
shale

gently dipping sequence of
SANDSTONE, SILTSTONE
and SHALE

taluvium and
landslide debris

river terrace
gravels

gently dipping sequence
of SANDSTONE,
SILTSTONE and SHALE

TEMPERATE CLIMATE

alluvium with
lenses of peat
and gravel in
valley with
permanent
river

*Figure 3.3 Temperate environments.*

## *Temperate environments (Figure 3.3)*

It is sometimes said rather light-heartedly that temperate landscapes are those left when the more distinctive hotter or colder environments of the world have been identified. In modern usage, the term temperate has come to be used to denote moderate climates with no great extremes of temperature. Temperate environments, which include some of the world's most densely populated areas, have presented some major design and construction challenges for engineers. These challenges have resulted in research that has been successful in temperate climates, but unsuccessful in other climatic regions where the geomorphological processes, environmental conditions and land use practices are different. This is an important point for engineers (Fookes *et al.*, 2005).

Today's temperate environment consists of five major subdivisions based on current and relict conditions. The first four subdivisions are forested, but these have been substantially altered by human activities over the last few thousand years. The fifth subdivision includes the steppes and prairies. One of the larger forested subdivisions is the Mediterranean environment (see Figure 3.4). The subdivisions are presented in Table 3.4.1.

### RELICT LANDFORMS

Present day temperate landscapes contain many relict landforms remaining from former climates, thus the majority of temperate features and their deposits probably do not now relate to contemporary environmental processes. Temperate environments can be thought of as domains of rain and rivers operating on a landscape that commonly contains the record of a variety of past environments. Domains, in this sense, represent equilibrium relationships between active processes related to controlling parameters, such as the climate, the infiltration rate, vegetation and land use.

The relict landscapes and local ground features that now lie within the temperate environment reflect the conditions that existed during the Quaternary. Four conceptual orders of change have been recognized.

- The first order of change involves large fluctuations in the water balance and temperature brought about by the changes from glacial to interglacial climates and vice versa. Each glacial/interglacial episode lasted for about 100,000 years and there were over 20 cycles during the important Pleistocene Period of the last two million or so years. The Pleistocene, together with the current geological age, the Holocene (about the last 10,000 years; alternatively called the Recent), make up the Quaternary. Holocene glaciers are now in retreat.

- The second order of change is caused by shorter climatic variations, such as glacial interstadials (i.e. minor retreats occurring within the major glacial episodes), each lasting a few hundred to a few thousand years.

- The third order of change is a result of changes in the local geomorphology (e.g. major earthquakes in tectonic zones) or human activity (e.g. deforestation).

- The fourth order of change is a result of events that may have persisted for only a few years (e.g. the effects of a major flood or landslide).

To simplify, existing temperate landforms can be considered as broadly composed of three major components.

- Extensive plateau areas, now often dissected, are the remains of landforms produced by pre-Quaternary erosion, e.g. planation surfaces formed under tropical or subtropical conditions, affected by later processes such as those occurring under glacial or periglacial conditions.

- Major valleys that were the product of erosion and deposition within the Quaternary as a result of glacial and interglacial episodes accompanied by major fluctuations in sea level.

- The detailed development of erosion and deposition along present river courses and valley floors during the current postglacial time period.

As many of the soils covering today's temperate areas are the relicts of former climatic or environmental conditions, it can be difficult to deduce how extensive a particular deposit is, and how much variability occurs, as it originated under conditions that no longer exist. Thus there is no simple basis for estimation in the development of the site geomodel. The range of deposits includes weathered material, colluvial, taluvial and fluvial materials originating from contemporary conditions, together with marine, aeolian, glacial, periglacial and tropical deposits that may still exist in significant amounts. The major problems associated with the interpretation of each type of deposit are indicated in Tables 3.3.1 and 3.4.1 and the Appendix.

Table 3.3.1 Types of deposit that may occur in temperate areas, including those developed in past climates (see also Figure 3.4 and the Appendix for discussion of individual soils).

| Type of deposit | Common characteristics | Associated engineering problems |
|---|---|---|
| Weathered and colluvial | Typically up to medium-grained fine soil size, not usually deep, may have developed on top of fossil deposits, podzolized soils now on surface | Some lateral variation as a result of subsurface changes in material |
| Fluvial (river) | Range of grain sizes present, often incorporates remnants of earlier deposits – frequent lateral and vertical changes | Small changes in river position can release new exposures of relict sediments |
| Marine | Range of grain sizes present, delivered by fluvial systems, may incorporate earlier deposits from cliff-falls or from offshore sediments | Deposits may also occur above the present sea level, marking former shorelines now uplifted, or made by former higher sea levels |
| Aeolian | Silts common in Vegetation Zone 5 (see Table 3.4.1) and also in other areas where vegetation and field boundaries have been removed; former conditions had loess or water-sorted loess ('brick earth' in Britain) cover or sand sheets, more common in Europe and northern Asia | May mantle surfaces and may not be related to deposits of a very different character beneath |
| *Relict deposits from former conditions* (see Figures 2.3, 2.4, 3.1, 3.2, 3.6, 3.7, 3.8 and 3.9) | | |
| Glacial | Till deposits, lake clays, glaciofluvial sands and gravels | Distribution and character not easily deduced; rapid variations in thickness |
| Periglacial | Angular scree deposits, unsorted slope deposits, fine wind-blown deposits; relict solifluction lobes and active layer slides | Distribution localized, but character reflects locally available rock and soil types – may occur on slopes that are unstable when modified by engineering |
| Tropical/ subtropical | Clays that are remnants of deep weathering (remains of Tertiary age climatic conditions) | Localized, often on plateau sites, may be locally deep |

[above centre] The power of water to erode and transport increases by the cube of its velocity.

[above right] Interlocking spurs form on opposite sides of a stream that meanders while it cuts deeper.

[left] Large meanders are developed by a river with a very low gradient on a wide floodplain, and are enlarged by erosion on the outside of the bends; the river cuts through meander necks to leave oxbow lakes; the insides of the meanders are filled with crescentic scrolls of sediment deposited in slack water.

[below left] Seen in the dry season, a moulin, or deep pothole, cut into bedrock by swirling water in a plunge pool at the foot of a waterfall; massive annual floods at this site scour the whole width of the bare rock.

[below right] Seen from the air, an alluvial fan is formed by a stream emerging from a steep gorge (here in deep shadow) and dropping its sediment load as it loses velocity over a wider and flatter basin floor.

Peat bogs dissected by small streams on a moorland plateau.

An old stone bridge that was largely destroyed when it was overtopped by a short-lived flood flow that was too large to pass through its arches.

A road has to lie along a river bank in a narrow valley, so has a gabion wall to prevent fluvial erosion undermining it during predictable high flood flows.

Alluvial sands and gravels in a floodplain quarry 10 m deep.

Gravel deposits on a slip-off slope inside the bend of a river.

A road undermined by multiple failures of hillside soils and artificial fill after a heavy rain storm event.

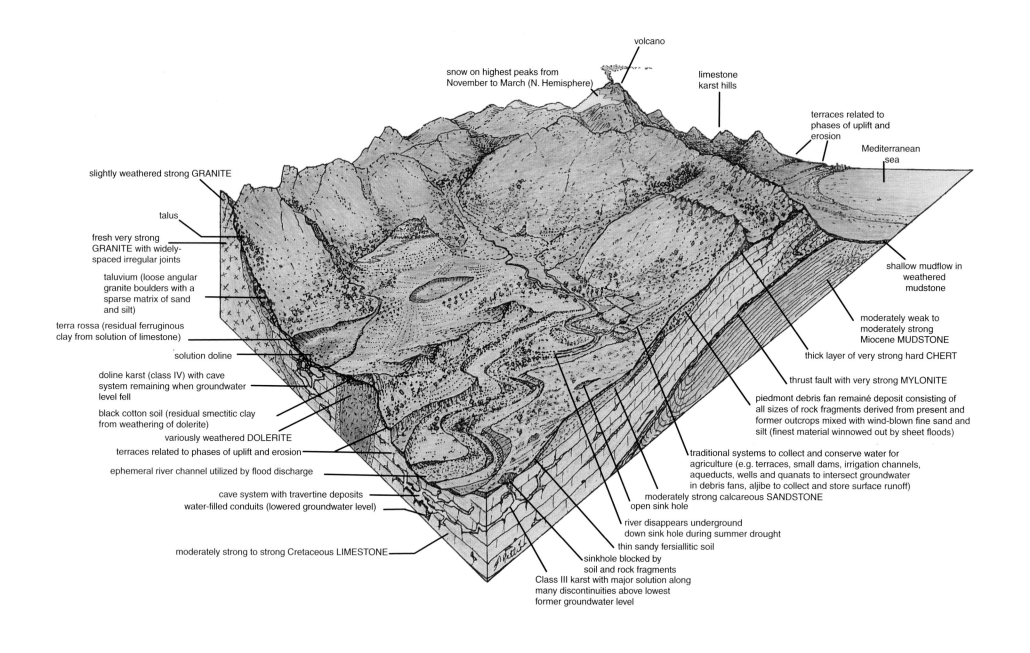

slightly weathered strong GRANITE

talus

fresh very strong
GRANITE with widely-
spaced irregular joints

taluvium (loose angular
granite boulders with a
sparse matrix of sand
and silt)

terra rossa (residual ferruginous
clay from solution of limestone)

solution doline

doline karst (class IV) with cave
system remaining when groundwater
level fell

black cotton soil (residual smectitic clay
from weathering of dolerite)

variously weathered DOLERITE

terraces related to phases of uplift and erosion

ephemeral river channel utilized by flood discharge

cave system with travertine deposits

water-filled conduits (lowered groundwater level)

moderately strong to strong Cretaceous LIMESTONE

volcano

snow on highest peaks from
November to March (N. Hemisphere)

limestone
karst hills

terraces related to
phases of uplift and
erosion

Mediterranean
sea

shallow mudflow in
weathered
mudstone

moderately weak to
moderately strong
Miocene MUDSTONE

thick layer of very strong hard CHERT

thrust fault with very strong MYLONITE

piedmont debris fan remainé deposit consisting of
all sizes of rock fragments derived from present and
former outcrops mixed with wind-blown fine sand and
silt (finest material winnowed out by sheet floods)

traditional systems to collect and conserve water for
agriculture (e.g. terraces, small dams, irrigation channels,
aqueducts, wells and quanats to intersect groundwater
in debris fans, aljibe to collect and store surface runoff)

moderately strong calcareous SANDSTONE

open sink hole

river disappears underground
down sink hole during summer drought

thin sandy fersiallitic soil

sinkhole blocked by
soil and rock fragments
Class III karst with major solution along
many discontinuities above lowest
former groundwater level

*Figure 3.4 Temperate environment: Mediterranean climate.*

## Temperate environments: the Mediterranean (*Figure 3.4*)

Figure 3.4 is based on land on the north shore of the Mediterranean Sea. However, the Mediterranean climate is characteristic of the western margins of continents in the world's warm temperate climates, typically between latitudes 30° and 40°. It is one of the five subdivisions of the temperate environment briefly discussed in Figure 3.3 and presented in Table 3.4.1. It is typified by hot, dry, sunny summers and warm, moist winters. In the summer the climate is dominated by subtropical anticyclones and in the winter by depressions. Within the Mediterranean basin itself, various temperate subdivisions are recognizable, depending on the distance from the Atlantic, but elsewhere in the world at this latitude the Mediterranean climate is restricted to narrower coastal margins commonly backed by topographic barriers, such as the coastal plain of central Chile backed by the Andes.

Mediterranean temperate areas adjacent to coasts also contain relict landforms of Quaternary climates (see Figure 3.3), notably anomalous buried channels or valleys. These were formed during glacial periods when the sea levels were lower than those of today because oceanic water was held in the continental ice sheets (see Introduction, Figures 3.1 and 3.2). Today these features may be steep-sided and flat-floored and may contain no drainage at all. Mediterranean environments (Zone 4 in Table 3.4.1) were once well forested, but have now largely been cleared by humans. Many landscapes have been changed significantly by various historical and modern forms of human activity.

Problems in the engineering characteristics of temperate deposits resulting from historical and current climate-related changes include the following.

- *A weathered ground profile.* A reduction in shear strength that may lead to slope failure, including failure on relict discontinuity planes; loss of bearing capacity; irregular bedrock profile.

Table 3.4.1 Subdivisions of the boreal forest temperate environments (modified after Tricart, 1957; Alexander, 1999). Under natural conditions Zones 1–4 would be forested, but these zones have been substantially transformed by human activity. Zones 1 and 5 are transitional to the other zones. This table applies to Figures 3.2, 3.3, 3.4, 3.9, 5.9 and Table 3.3.1.

| Vegetation Zone | Climate | Processes and features | Major hazards |
|---|---|---|---|
| 1: Forest on old Pleistocene permafrost (deciduous trees such as larch, birch and aspen) | Severe winters, may be associated with high-altitude periglacial areas | Transitional to the modern periglacial zone, local permanently frozen ground beneath the land surface may be continuous or discontinuous and is residual from the Pleistocene and not forming at present, hence 'temperate' | Wildfires, frost or ice storms, snowstorms, wind storms; modification of ground surface with attendant problems |
| 2: Forested zone of middle latitudes, mild (deciduous broadleaf woodland) | Maritime without severe winters; no large seasonal variations in temperature or humidity | Chemical erosion limited by moderate temperatures, some frost action, but penetration rarely reaches bedrock; high-angle slopes can be stable where still covered by forest. Landslides on devegetated slopes. Flooding may increase downstream of vegetation changes. | Accelerated erosion; frost heave and collapse; floods; landslides |
| 3: Forested zone of middle latitudes, cold (evergreen conifers such as spruce, fir and pine) | Severe winters and seasonally distributed precipitation | Heavy showers and snowmelt can produce higher peak streamflow rates than in Zone 2; mechanical processes more important as frost penetration is great and can reach bedrock; slopes can be steep, up to 20° to 35°, and covered in slope debris; chemical erosion limited by winter frost | Drought; severe thunderstorms; hailstorms; snowstorms; landslides, especially when vegetation removed; downstream flooding increases after vegetation changes |
| 4: Mediterranean forested zone of middle latitudes (evergreen oaks, pine) | Seasonal precipitation, mild winters, warm/hot summers; frost uncommon at low elevations | Alternations of wet and dry conditions exert major influences, such as inducing landslides; seasonal streamflow regime can give high seasonal discharges that transport coarse debris; rapid dissection and gullying where vegetation removed or degraded | Soil erosion; floods with high spatial and temporal variability; high river sediment yields; landslides, sheet erosion where vegetation removed; increased flooding downstream |
| 5: Sub-desert steppes and prairies | Summer rainstorms, dry cold winters | Transitional to savannas and deserts with some frost action in winter; wind action, occasional sheet wash and gullying | Drought; tornadoes; soil erosion. Deflation and gullying encouraged by removal of vegetation. |

- *Expansive soils.* Often, but not always, a residual weathered soil. Swelling/shrinkage with moisture changes, especially in areas where smectite clays have developed.

- *Consolidation.* Differential amounts of settlement as a result of variations within relict and bedrock deposits and in their stratigraphy.

- *Drainage.* Settlement of deposits that can give substantial, commonly irregular, lowering of surface elevation and so increase the risk of flooding. Formation of desiccation cracks that may provide locations for slumps or gulley development.

Problems resulting from human activities include the following.

- *Groundwater.* Groundwater rise due to the creation of reservoirs, with a consequent increase in pore pressure and slope failure.

- *Piping.* Land use practices (including deforestation, over-use and poor management of agricultural land and population pressure) that lead to surface erosion, subsurface piping and the possibility of collapse and gully development.

The characteristics of soils found within temperate environments are discussed further in the Appendix.

Within the Mediterranean terrains, strong limestones form many of the bare crags and hill features because they are more resistant to erosion in the mild pluvial conditions.

A deep and straight U-shaped valley is a relict of Pleistocene glaciations, features from which are characteristic of the Mediterranean environments.

Wide river terraces, formed of alluvial gravels, sands silts and clays, are especially common in areas that were subjected to episodic uplift in Alpine orogenies.

[left] In pluvial terrains, gully erosion causes long-term dissection of sloping ground, where streams cut down through old unconsolidated slope sediments and into weaker bedrock beneath; remnants of the earlier slope profiles form benches that are suitable for cultivation.

[below] The end-product of gully erosion in weak and cohesive rocks can be a landscape of tall pinnacles.

Short-term gully erosion can be extremely destructive of good agicultural land on steep slopes, sometimes with major features developing during single storm events.

[left] Loess is one of the materials most prone to piping; cavities are created along sub-surface drainage routes by mechanical removal of first the fines and then headward ersoion that carries the larger soil particles out through the downstream end of the expanding pipe.

[below] Sinkhole collapse at the top end of a soil pipe.

A limestone plateau with a valley cut largely by Pleistocene periglacial erosion.

On the larger limestone plateaus, the extensive dendritic systems of dry valleys were largely formed by stream erosion when the ground was frozen and impermeable during cold stages of the Pleistocene.

[above] Subsidence sinkholes can form in any soil that covers cavernous rock, in this case gypsum, and are notably frequent after rain events on agricultural land where the natural drainage has been modified.

[right] A river at its normal level in a period of stable weather, and at flood level where it is constrained by an artificial levee built to protect an urban area.

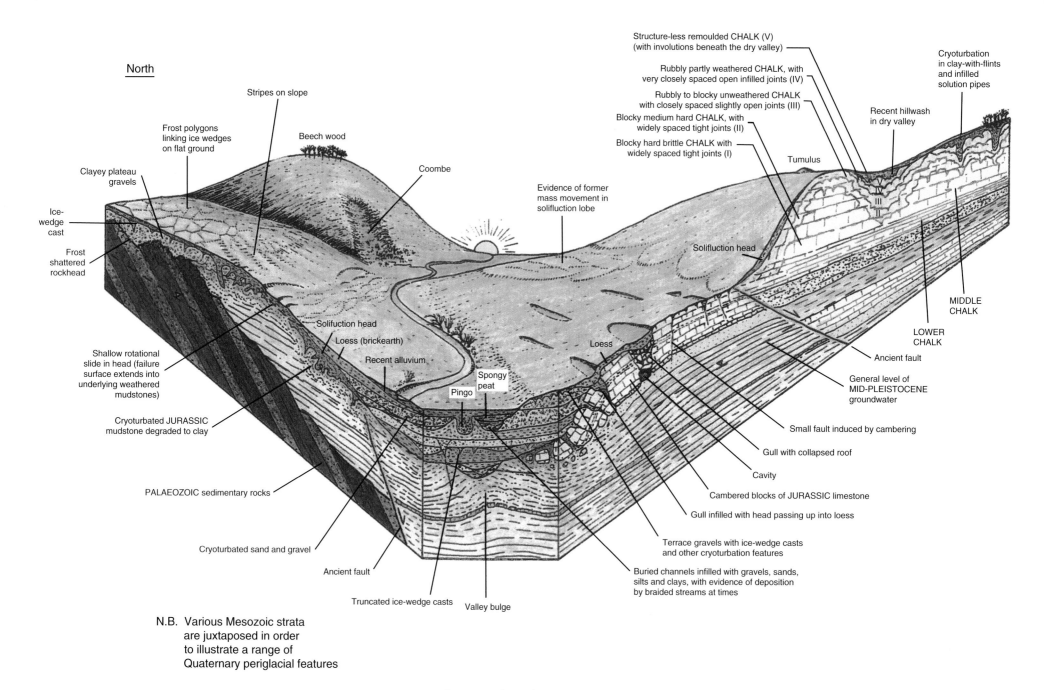

North

Stripes on slope

Frost polygons
linking ice wedges
on flat ground

Beech wood

Clayey plateau
gravels

Coombe

Structure-less remoulded CHALK (V)
(with involutions beneath the dry valley)

Rubbly partly weathered CHALK, with
very closely spaced open infilled joints (IV)

Rubbly to blocky unweathered CHALK
with closely spaced slightly open joints (III)

Blocky medium hard CHALK, with
widely spaced tight joints (II)

Blocky hard brittle CHALK with
widely spaced tight joints (I)

Cryoturbation
in clay-with-flints
and infilled
solution pipes

Recent hillwash
in dry valley

Tumulus

Ice-
wedge
cast

Evidence of former
mass movement in
solifluction lobe

Frost
shattered
rockhead

Solifluction head

Shallow rotational
slide in head (failure
surface extends into
underlying weathered
mudstones)

Solifuction head

Loess (brickearth)

Recent alluvium

MIDDLE
CHALK

LOWER
CHALK

Ancient fault

Loess

Cryoturbated JURASSIC
mudstone degraded to clay

Pingo

Spongy
peat

General level of
MID-PLEISTOCENE
groundwater

Small fault induced by cambering

PALAEOZOIC sedimentary rocks

Gull with collapsed roof

Cavity

Cambered blocks of JURASSIC limestone

Gull infilled with head passing up into loess

Cryoturbated sand and gravel

Terrace gravels with ice-wedge casts
and other cryoturbation features

Ancient fault

Buried channels infilled with gravels, sands,
silts and clays, with evidence of deposition
by braided streams at times

Truncated ice-wedge casts

Valley bulge

N.B.  Various Mesozoic strata
are juxtaposed in order
to illustrate a range of
Quaternary periglacial features

*Figure 3.5 A relict periglacial terrain: southern Britain.*

## *Relict periglacial terrain: southern Britain (Figure 3.5)*

In vast areas that are no longer subject to periglacial conditions, the observed surface and subsurface forms and soil characteristics commonly reflect former periglacial processes and forms. Evidence of these forms includes involutions, solifluction, valley bulging, cambering, frost mounds, peat deposits, pingo craters, ice-wedge casts and crop (ice-wedge) polygons (see Figures 3.2 and 5.9). Hutchinson (1991) writes that, in Britain, 'the reactivation of relic clayey solifluction mantles by ill-advised earthworks probably constitutes at present the most frequent and costly type of failure … having a periglacial origin'. Solifluction mantles (commonly called head) typically contain pre-existing shear planes that are easily missed in logging borehole cores. Pits must be dug to investigate fully and to develop the site model. Modern slope failures commonly utilize the shear planes.

Although Figure 3.5 has been modelled on southern Britain, all of the relict features can be found elsewhere in areas of former periglacial activity, especially across large parts of northern Britain, North America, Siberia and north-western Europe. Air photography reveals many periglacial forms and patterns throughout these areas – for example, several hundred fossil ice wedges have been identified from the air in Denmark, with more in the Netherlands and Poland.

These relict periglacial features are of significant concern to civil engineering because they may represent sudden and usually unexpected replacement of one material by another, perhaps with inferior geotechnical properties (Table 3.5.1). Relict features are especially important for shallow foundations, roads, runways and canals. Near-surface features invariably make interpretation by cable and tool rigs (commonly referred to, in Britain, as 'shell-and-auger') and similar light percussion boring techniques difficult or impossible. Pits and trenches are needed to produce more reliable interpretations and samples.

### Loess

Many of the former periglaciated areas are covered in wind-blown silt (loess) that was derived from the outwash plains of

Table 3.5.1 Examples of geomorphological problems found in former periglacial terrains.

| Problem | Effects | Examples |
|---|---|---|
| *Near-surface disturbances* | | |
| Involutions (cryoturbations) | Surface sediment diversity and disruption | Common over much of Britain and elsewhere |
| Frost-shattering | Weakened rock profiles, liable to failure; scree (talus)-covered slopes | Near-surface features on rocks such as chalk |
| Solifluction | Reactivation of slip surfaces below solifluction lobes | Mass movements during road construction, e.g. major problems in Sevenoaks bypass, Kent, UK |
| Ice wedges | Fine material filling cracks likely to pose ground stability problems | Collapse of trench walls during, for example, gas pipe laying, differential settlement |
| *Deeper seated disturbances* | | |
| Cambering and valley bulging | Down-slope movement of competent rock over weaker, more ductile rocks; formation of large cracks ('gulls') | Widespread in limestones and sandstones that overlie clays of (for example) south-eastern Britain; gulls can be a serious problem in new housing developments |
| Fossil pingos | Shallow craters buried by sand and gravel; common in lowlands where groundwater froze during periglacial conditions | Cause of serious foundation problems, e.g. Battersea power station, London, UK |
| *Loess and cover sands* | | |
| Deep loess | Lower density, high porosity, highly permeable metastable material, often hundreds of metres thick (not in Britain), prone to sliding and surface gullying | Major problems of slope instability in urban areas, e.g. notably Gansu, China; Germany; Poland |
| Cover sands | Thinner, but can be easily eroded, may be metastable | Fine silt eroded from thin cover sand deposits, e.g. affects urban water supply treatment in southern Pennines, UK |

retreating glaciers. These may extend far beyond the areas of Pleistocene permafrost. Loess landforms vary from extensive plateaus to small hills and are frequently cut by deep, vertical-sided gullies; vertical fissures within the loess dominate its permeability characteristics. Areas with thick layers of loess are subject to various forms of landslides and erosion. Figure 3.5 shows a small amount of thin loess (commonly called brick earth in Britain) on a shallow valley side. Huge areas with loess mantles up to tens of metres thick occur in the USA and across northern Europe through Russia to China, but do not occur in Britain (Derbyshire and Meng, 1995).

The dominant mineral in most of the world's loess is quartz. Significant amounts of feldspars and mica may also be present and minor carbonate and clay minerals may locally constitute a cement. Because the quartz grains are deposited by air-fall, they are typically loosely packed, moderate to poorly sorted

and lack stratification. The ambient natural moisture of loess is broadly controlled by the local climate regime. In general, the shear strength of loess increases when the moisture content decreases and there is an increase in bulk density. The bearing capacity increases consistently with depth, except that higher values occur locally in clay-enriched materials (i.e. palaeosols, ancient buried soils).

When loess is put under a load (e.g. from engineering constructions), it consolidates in the normal manner for a silt with a high void ratio, but, if wetted under the new load, it will suddenly collapse when close to saturation. This potentially dangerous metastable behaviour should be investigated by laboratory tests (e.g. a flooded oedometer) before construction. Hydrocompaction occurs when porous loess, wetted naturally, collapses.

CHALK

Figure 3.5 shows faulted Cretaceous chalk (see Table 3.5.2) overlying various Jurassic sediments, common in southern Britain. Chalk also occurs extensively across Europe and North America. It is a weak, fine-grained, white or grey porous limestone of marine origin formed by pelagic shells deposited in a tropical shelf sea in Cretaceous times. It varies in hardness and strength; some 'hard' beds, each up to about 2 m thick, occur within the chalk sequence, which totals several hundred metres. The chalk outcrop normally forms a prominent escarpment with the hard bands forming inter- mediate escarpments. Chalk in England generally gives rise to distinctive rolling country known as 'the Downs'. It typically has a trellis pattern of drainage with valleys that are mostly dry or which may contain ephemeral streams (i.e. 'winter bournes') and coombes (bowl-shaped hollows that occur on the flanks of dry valleys). These were developed when the landscape was frozen during the Pleistocene and were cut by water during the summer melts.

During periglacial conditions, chalk was typically weathered by freeze–thaw action (including frost-shattering) down to several tens of metres (see Table 5.1.1 and discussion in Figure 5.1). This has lent itself to a classification of chalk weathering from fresh chalk (Grade I) to a structure-less remoulded chalk (Grade V). Other common periglacial con- ditions include dissolution features, sink holes and small cave systems. These are commonly back-filled with loess, fine sand and sludge deposits. Streams flowing on impermeable Tertiary sediments (or glacial till) often disappear down swallow holes on encountering the chalk, e.g. at Farnham, Surrey, UK. Chalk generally provides a satisfactory foundation material, but investigation must evaluate the presence or otherwise of periglacial features that are important in engineering, e.g. sink holes (see Figure 3.5; Bell and Culshaw, 2005).

See also Figure 5.1 case history on periglacial areas of southern Britain.

Table 3.5.2 Basic English Chalk stratigraphy (after Rawson *et al.*, 2001 and Mortimore et al., 2001).

| Old units | Stage | Sub-group | Southern area | | Northern area | |
|---|---|---|---|---|---|---|
| | | | Formation | Member | Formation | Member |
| Upper Chalk | Campanian | White Chalk | Portsdown Chalk | | Rowe Chalk | |
| | | | Culver Chalk | Spetisbury Chalk | Flamborough Chalk | |
| | | | | Tarrant Chalk | | |
| | | | Newhaven Chalk | | | |
| | Santonian | | Seaford Chalk | | Burnham Chalk | |
| | Coniacian | | | | | |
| | | | Lewes Nodular Chalk | | | |
| | Turonian | | New Pit Chalk | | Welton Chalk | |
| Middle Chalk | | | Holywell Nodular Chalk | Plenus Marls | | Plenus Marls |
| Lower Chalk | Cenomanian | Grey Chalk | Zig/Zag Chalk | | Ferriby Chalk | |
| | | | West Melbury Marly Chalk | | | |

A low ridge around a depression is the former cover of soil that slipped off a pingo, which then melted away; so the pingo hill became an 'ognip' hollow.

Many large landslides were initiated during periglacial times, but are features of de-glaciation where over-steepened hillsides failed due to loss of support by ice that had filled the valleys.

Ramparts of talus (scree) of frost-shattered debris, formed largely in periglacial conditions, are less active and grass over in the current warmer climate.

Tors of granite in situ, with residual blocks that were probably exposed in a periglacial environment after deep weathering down joints in warmer climates.

[above] The thin soil over a limestone plateau in the Pennines is largely formed of periglacial loess blown in from outwash plains in the Irish Sea basin.

[below] Loess in Britain is minimal compared with the hugely thick and extensive loess in central Asia.

A raised beach, and an old sea cliff behind it, formed before the depressed land rose by isostatic uplift when the loading of a Quaternary ice sheet was lost by melting.

[left] Sands and gravels forming a raised beach on an old wave-cut platform are typically variable, both laterally and vertically.

A camber fold over soft clay has open gulls in limestone that is sliding towards an adjacent valley off to the right.

[above] A quarried section exposes about two metres of broken limestone that was frost shattered in periglacial conditions of the Quaternary, beneath which fresh rock forms a stable roof over a gallery in an old stone mine.

[above left] Terracettes on a limestone hillside have been formed by long-term creep of the thin soil cover.

[left] A coastal cliff exposes frost-shattered chalk beside a buried valley filled with inclined layers of soliflucted 'coombe rock' of rubbly chalk debris that accumulated under the periglacial conditions of the Quaternary.

[below left] Dry valleys, coombes, formed by surface drainage and solifluction when underlying chalk was rendered impermeable by its permafrost ground ice.

A cryoturbated interface between chalk and a residual clay soil formed by periglacial ground ice heave that has been complicated by later dissolution of the chalk.

'Rubble chalk', the weak, angular rock debris that was formed by weathering and frost action reaching as much as 10 metres deep under periglacial conditions during cold Quaternary stages.

A doline in soil-covered chalk formed by dissolution of ground originally disturbed by periglacial permafrost.

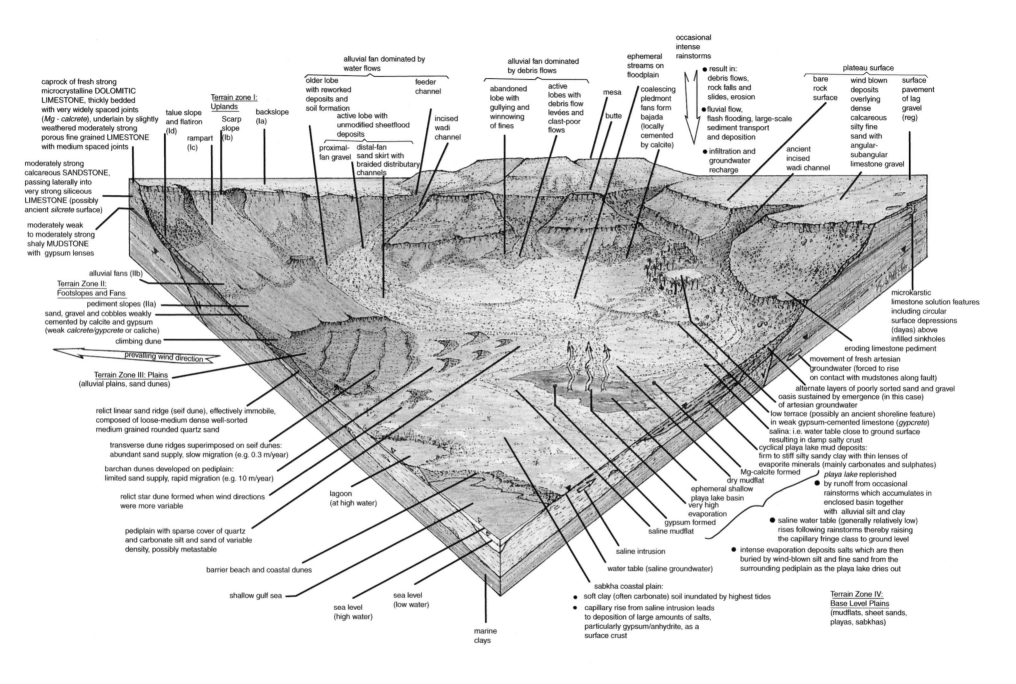

*Figure 3.6 Hot desert environments: drylands.*

98

## Hot desert environments: drylands (Figure 3.6)

Hot drylands cover about one-third of the Earth's surface and can be distinguished through their aridity index P/ETP, where P is the annual precipitation and ETP is the mean annual potential evapotranspiration. As the aridity increases, the P/ETP ratio decreases. UNESCO (1979) defines four zones of aridity: *sub-humid, semi-arid, arid* and *hyper-arid*, with P/ETP values of 0.50–0.75, 0.20–0.50, 0.03–0.20 and <0.03, respectively.

Most hot drylands are centred on the tropics where the stable descending air of the Subtropical High Pressure Zone maintains aridity throughout the year (see Figures 1.1 and 1.2). Another important factor controlling the distribution of arid zones is the presence of large land masses that disrupt the zonal pattern of global pressure systems (e.g. the Sahara Desert). Mountain barriers generate rain shadows (e.g. the Mohave Desert) and cold oceans bordering hot lands prevent condensation and rainfall from inshore winds (e.g. the Atacama Desert); the same happens on the eastern margins of warm oceans (e.g. the Namib Desert) (Lee and Fookes, 2005; Walker, 2012).

The amount of rainfall displays extreme spatial variability and local variations in intensity can be considerable; within the same storm some areas can receive 20 times more rainfall than sites a few kilometres away. Cloudbursts are a characteristic feature of this climate type and are associated with almost instantaneous peaks in flood flows (flash floods) in stream channels (wadis, arroyos), typically followed by a long tail of low intensity rain.

The popular notion is that barren hot drylands are flat and dominated by sand dunes. However, the reality is rather different and much more complex. There are two main types of desert landscape: *shield and platform deserts*, including inselberg–pediment landscapes and canyon–scarp–pediment landscapes, and *basin and range* deserts.

Much of the detail of current desert landscapes is the product of the major climatic changes experienced throughout the Quaternary when there were many phases of alternat-ing semi-arid or humid and arid conditions. However, the important characteristic of desert environments is the marked contrast between the long-term stability of many former and current upland landforms, including escarpments and pediments, and the current and past dynamic geomorphological activity of lowland landforms, such as alluvial plains, sand seas and alluvial fans.

Most desert soils are granular and the grading is broadly related to the distance from the upland sources: coarser sediments closer to the mountains are moved by water and finer material further away is transported by water or wind. Most fine sediments are stored in topographic lows or on base-level plains, where the high water-tables with saline groundwater can create engineering problems as a result of salt weathering. Potentially metastable wind-blown dust (loess) accumulated across many desert landscapes during the Quaternary.

For engineering geomorphological purposes the key to hot dryland issues can be related to a number of distinct terrain units characteristic of desert environments, albeit in varying combinations. Four main zones can be recognized from the uplands down to the base-level plains (the level below which no further erosion will occur), each with typical surface features and with different engineering behaviours. Potential engineering problems encountered in drylands include: surface erosion and instability; difficult excavation, especially in areas with a near-surface hardening of soils and rocks to form duricrusts (see Figure 2.5); the granular behaviour of desert soils; the availability of construction aggregates; water and sediment movement problems, including flash floods; wind-blown sand; and aggressive salty ground (see extensive discussion in Walker, 2012).

- *Zone I: the uplands.* Where mechanical weathering is dominant, the area is commonly characterized by bare rock and boulder-strewn slopes. Subzones include back slopes, scarp slopes and free faces, ramparts, talus slopes and flat irons.
- *Zone II: foot slopes and fans.* The terrain surrounding the uplands may consist of two contrasting terrain units: rock pediments and extensive coarse alluvial fans. Relict forms may also be present. Pediments are gently sloping surfaces developed in the bedrock and commonly cutting across a range of rock types. Alluvial fans are complex cones of poorly sorted sediments laid down where a channel emerged from an upland area onto a plain; the bed load is spread out over a growing fan during rainstorms. Routing across fans therefore has to be investigated and planned carefully. This zone provides construction materials for roads and concrete.
- *Zone III: plains.* This is normally a wide zone with a net deposition of sediments supplied by erosion of the adjacent upland and foot slopes. It typically has three subzones: alluvial sediments, wind-blown sediments and stone-covered rocky surfaces. These very gentle slopes may overlie a considerable thickness of water-borne material. Fluvial and aeolian processes generate the main geohazards and dominate engineering issues. Sand for construction purposes is supplied by this zone.
- *Zone IV: base-level plains.* This zone is characterized by the effects of near-surface groundwater and salt accumulation, usually in soils dominated by wind-blown and water-deposited silts and sands. Coastal sabkhas and inland playas – salt-rich areas aggressive to engineering – commonly occur. Two subzones are recognized: enclosed basins and broad depressions. Fine-grained soils in areas of highly saline groundwater have restricted load-bearing and other engineering performance characteristics. Excavations may need to be dewatered, with filter protection against the migration of fine-grained particles.

Calcrete duricrusts (e.g. carbonate-cemented sands) may be present in layered sediments, especially in coastal sabkhas. The local salt regime may be complex and is likely to vary seasonally. Salts will contaminate fine and coarse aggregates and lead to salt attacks on roads, structures and buildings. Each site needs to be investigated and then sampled and tested individually.

A rocky desert surface is stripped down to a layer of hard rock, with sediment only in shallow wadis; remnants of a flat, earlier, higher desert surface form the tops of mesas in the distance.

An incised wadi in Zone II carries occasional flash floods that transport and deposit large amounts of coarse, clastic debris; the bank on the right exposes similar stratified debris, where the wadi was cut deeper by erosion in a wetter climate a few thousand years ago.

[above] A fluvial slot canyon cut in desert rock by flash floods from rainstorms on distant mountains.

[right] Linear dunes, formed by two seasonal wind systems, are spaced about 800 metres apart and extend for tens of kilometres in Zone IV.

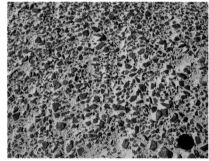

Giant star dunes of blown sand, over 100 metres tall, were mostly formed during the Quaternary by faster winds from variable directions (Zone III).

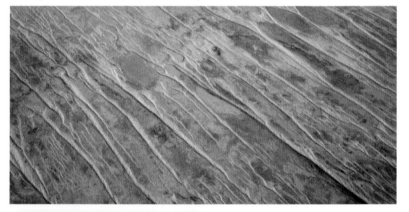

Desert pavement evolves as a layer of gravel-sized stones: [far left] wind-polished pebbles, known as ventifacts, still have sand remaining between them; [near left] all the sand has been removed by the wind, and only pebbles remain to create the armoured desert surface.

Black disc 70mm across; both images.

[left] Barchans are small sand dunes that migrate a few metres per year, here from right to left, due to wind from a single direction in Zone III.

[right] The side of a single large barchan that has swept over a road, necessitating its re-routing.

Playa floor (Zone IV) of dry mud on lake bed in a desert basin; classic profile of footslope and steeper uplands (II/I) on skyline.

Desert basin lake, with salt forming the floor of the shallows and the distant flats beyond low banks of gypsum and clay.

Polygonal pressure ridges in a salt floor, formed over cracks during desiccation phases (Zone IV).

Desert road conditions: [from left] transverse corrugations on sand; flash flood through a desert town; flood erosion of a road embankment along a wadi, and a new track up its floor.

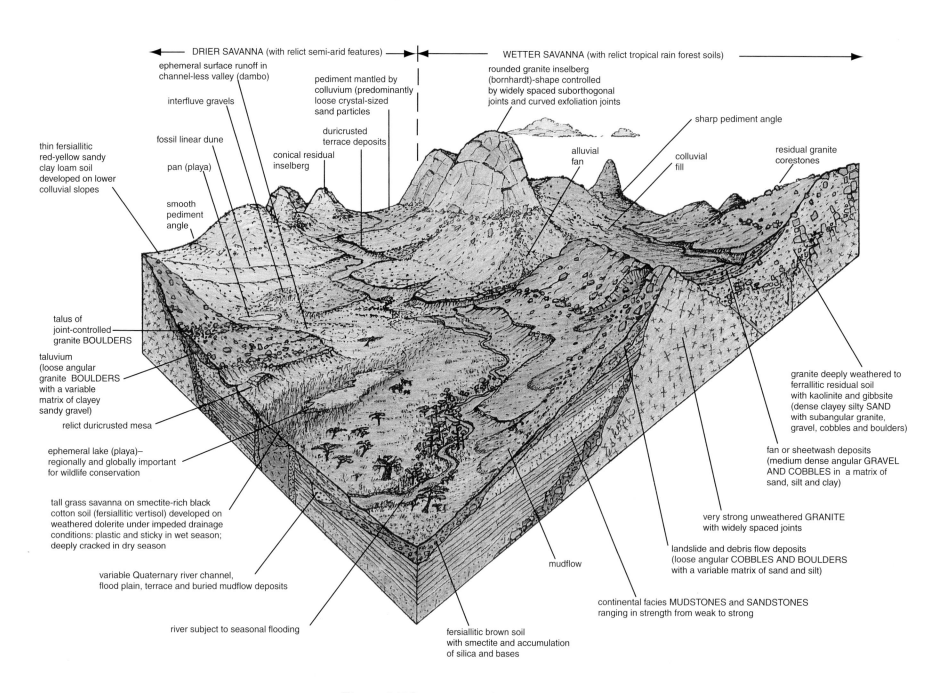

DRIER SAVANNA (with relict semi-arid features)

WETTER SAVANNA (with relict tropical rain forest soils)

ephemeral surface runoff in
channel-less valley (dambo)

pediment mantled by
colluvium (predominantly
loose crystal-sized
sand particles

rounded granite inselberg
(bornhardt)-shape controlled
by widely spaced suborthogonal
joints and curved exfoliation joints

interfluve gravels

sharp pediment angle

duricrusted
terrace deposits

thin fersiallitic
red-yellow sandy
clay loam soil
developed on lower
colluvial slopes

fossil linear dune

alluvial
fan

colluvial
fill

residual granite
corestones

conical residual
inselberg

pan (playa)

smooth
pediment
angle

talus of
joint-controlled
granite BOULDERS

taluvium
(loose angular
granite BOULDERS
with a variable
matrix of clayey
sandy gravel)

granite deeply weathered to
ferrallitic residual soil
with kaolinite and gibbsite
(dense clayey silty SAND
with subangular granite,
gravel, cobbles and boulders)

relict duricrusted mesa

fan or sheetwash deposits
(medium dense angular GRAVEL
AND COBBLES in a matrix of
sand, silt and clay)

ephemeral lake (playa)—
regionally and globally important
for wildlife conservation

very strong unweathered GRANITE
with widely spaced joints

tall grass savanna on smectite-rich black
cotton soil (fersiallitic vertisol) developed on
weathered dolerite under impeded drainage
conditions: plastic and sticky in wet season;
deeply cracked in dry season

landslide and debris flow deposits
(loose angular COBBLES AND BOULDERS
with a variable matrix of sand and silt)

variable Quaternary river channel,
flood plain, terrace and buried mudflow deposits

mudflow

continental facies MUDSTONES and SANDSTONES
ranging in strength from weak to strong

river subject to seasonal flooding

fersiallitic brown soil
with smectite and accumulation
of silica and bases

*Figure 3.7 Savanna environments.*

102

## Savanna environments (Figure 3.7)

Savanna terrains are generally regarded as environments of the seasonal tropics (i.e. annual wet and relatively dry seasons) that lie beyond the range of wet (humid) tropical forests, but without the extremes of deserts and steppes. Present day savanna vegetation consists of a variety of open deciduous woodland, woodland and grassland mosaics, and areas of open grassland. The transition to tropical rain forest occurs when the amount of rainfall approaches about 1500 mm/year and the dry season is less than four months long. When the dry season conditions become more extreme and the rainfall decreases to <80 mm/year, the deciduous broadleaved species give way to acacia thorn woodland and semi-arid steppe.

The largest areas of savanna coincide with the extensive plateau (platform) surfaces of the ancient southern Gondwana continents, planated in the late Mesozoic or early Cenozoic Eras. The savannas are largely cratonic areas underlain by ancient Archaean rocks and are typically covered by younger undeformed platform sediments. These ancient land surfaces are dominated by poor leached soils, commonly underlain by deep saprolites (see Figures 2.2–2.4), which may be capped by duricrusts (see Figure 2.5). They are also characterized by groups of inselbergs, typically prominent rocky hills of granite that are domed in profile. Important areas of basic volcanic rocks can also occur, notably the Deccan Plateau of India and the area adjacent to the East African rift zone.

The climatic history has created many complexities in the ground profile of savanna landscapes, particularly where prolonged weathering has reached great depths, in places more than 100 m deep. Importantly, the geotechnical characteristics of this weathering mantle may vary greatly depending on the parent rocks and the complex hydrological history. The soil mantle is often characterized by a widespread cover of surface or near-surface resistant duricrusts. These duricrusts were first formed in the saprolites or near-surface sediments by the precipitation of hydrated oxides of aluminium, iron and silicon, or sometimes as calcium carbonate or calcium sulphate, often followed by modification. The formation of the various types of duricrust materials generally follows the humidity of recent past climates as they varied between wet and dry conditions. For further discussion on the complexity of savanna duricrusts, see Thomas (2005).

Many hill-slopes have strong catenary relationships (repeated sequences of soil profiles), which developed in marked seasonal wet and dry climates, and related varying groundwater conditions (Figure 3.7; see also Figure 2.5 and Fookes, 1997).

### IMPLICATIONS FOR FOUNDATIONS AND EARTHWORKS

The conditions for shallow foundations are reasonably good in most savanna terrains. The allowable bearing capacities tend to decrease with increasing clay content. Problems that may be encountered include the excessive settlement of collapsible soils or heave in expansive soils. Useful introductions to these soils have been provided by the American Society of Civil Engineers (1982) and Blight (1997).

Undermining by erosion beneath duricrusts leads to the possibility of local ground subsidence or building settlement. Piling for heavy structures can be problematic as a result of the variability of the ground characteristics, both horizontally and vertically, with the potential for zones of fairly strong materials to be underlain by more weathered and/or weaker ground. Laterites form a large part of the upper weathered profile and consist of gravelly varieties commonly used as construction materials for roads (see Figure 2.5; Walker, 2012).

During the wet season, tropical savannas share rainfall characteristics with more humid zones, but experience precipitation for a shorter period of time, generally less than six months of the year. Rainfall often occurs as intense downpours, perhaps delivering more than 100 mm in a single storm lasting two to three hours. In drier areas, a characteristic monsoon climate occurs (one annual windy and very wet period several months long). Dry periods often produce fires. The restricted amount of groundwater across much of the savanna environment appears to limit the potential for deep-seated landslides. Nevertheless, the impact of large, intense storms can be devas-tating on steeper hill-slopes, leading to numerous small debris slides that can translate into mudflows down-slope.

Gully erosion is common on deep colluvium, often triggered by cattle tracks and by uncontrolled run-off from roads. Once gully incision has taken place, the hydraulic gradient is greatly increased at the head-cut and groundwater penetration through cracks is rapid. Piping (erosion by underground water) occurs, enlarging cavities until roof collapses become common.

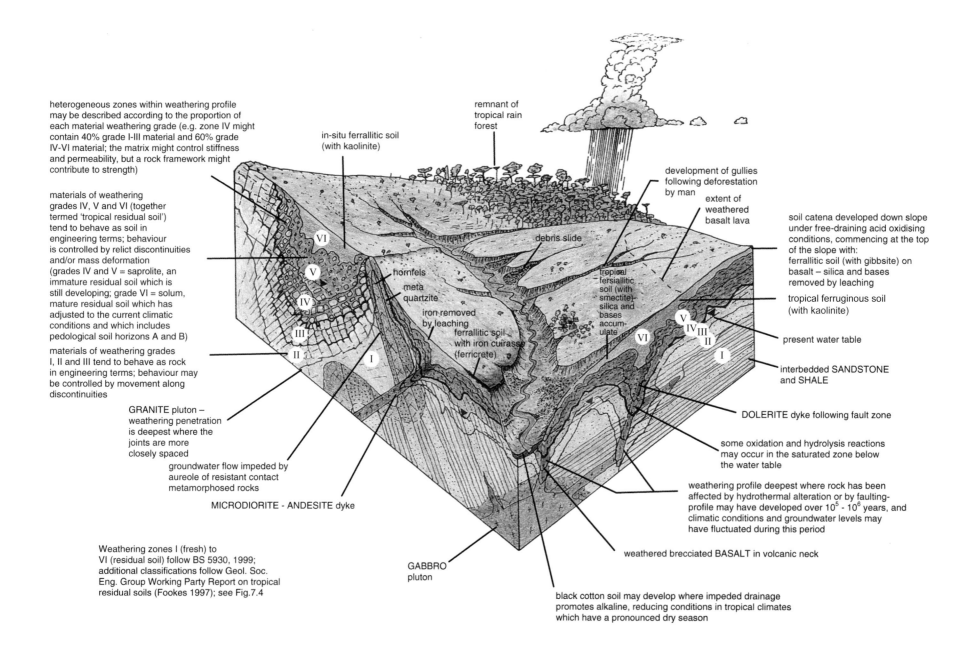

heterogeneous zones within weathering profile may be described according to the proportion of each material weathering grade (e.g. zone IV might contain 40% grade I-III material and 60% grade IV-VI material; the matrix might control stiffness and permeability, but a rock framework might contribute to strength)

materials of weathering grades IV, V and VI (together termed 'tropical residual soil') tend to behave as soil in engineering terms; behaviour is controlled by relict discontinuities and/or mass deformation (grades IV and V = saprolite, an immature residual soil which is still developing; grade VI = solum, mature residual soil which has adjusted to the current climatic conditions and which includes pedological soil horizons A and B)

materials of weathering grades I, II and III tend to behave as rock in engineering terms; behaviour may be controlled by movement along discontinuities

GRANITE pluton – weathering penetration is deepest where the joints are more closely spaced

groundwater flow impeded by aureole of resistant contact metamorphosed rocks

MICRODIORITE - ANDESITE dyke

Weathering zones I (fresh) to VI (residual soil) follow BS 5930, 1999; additional classifications follow Geol. Soc. Eng. Group Working Party Report on tropical residual soils (Fookes 1997); see Fig.7.4

in-situ ferrallitic soil (with kaolinite)

remnant of tropical rain forest

debris slide

hornfels

meta quartzite

iron removed by leaching ferrallitic soil with iron cuirasse (ferricrete)

tropical fersiallitic soil (with smectite)- silica and bases accumulate

development of gullies following deforestation by man

extent of weathered basalt lava

soil catena developed down slope under free-draining acid oxidising conditions, commencing at the top of the slope with: ferrallitic soil (with gibbsite) on basalt – silica and bases removed by leaching

tropical ferruginous soil (with kaolinite)

present water table

interbedded SANDSTONE and SHALE

DOLERITE dyke following fault zone

some oxidation and hydrolysis reactions may occur in the saturated zone below the water table

weathering profile deepest where rock has been affected by hydrothermal alteration or by faulting- profile may have developed over $10^5$ - $10^6$ years, and climatic conditions and groundwater levels may have fluctuated during this period

weathered brecciated BASALT in volcanic neck

GABBRO pluton

black cotton soil may develop where impeded drainage promotes alkaline, reducing conditions in tropical climates which have a pronounced dry season

*Figure 3.8 Hot wet tropical environments.*

## Hot wet tropical environments (Figure 3.8)

The term 'hot wet tropical' is used here to include all hot, high rainfall regions. These are sometimes called the 'hot wetlands'. The high temperature and humidity of wet tropical lands support a dense, diverse forest cover and continuous biological activity, commonly called 'tropical rain forest'. The apparent uniformity of the humid tropical rain forest is deceptive. There are wide variations in forest type (Table 3.8.1) and in the associated engineering soils, rocks and landforms beneath the forest. This diversity must be evaluated individually because engineering experience in one part of this region cannot necessarily be translated to another (Douglas, 2005; see Figure 2.2).

The wet tropics embrace diverse landscapes: the tectonically active mountains of the equatorial parts of the Pacific Rim of Fire; the volcanic landscapes of Java and central America; the old plateau of the northern part of the Brazilian Shield and the flooded forests and freshwater swamp forests of the lower parts of the Amazon and Congo basins, eastern Sumatra and southern Borneo. The tectonically active areas tend to have younger rocks, which weather and break up more rapidly than the older and stronger rocks of the ancient, more stable areas. Earthquakes often trigger landslides, which help to remove near-surface decomposing rocks and supply large amounts of sediments to rivers (see Figure 3.9).

The rates of chemical weathering in the wet tropics are high, leading to the development of residual soils (see Figures 2.3, 2.4, 2.5 and Table 3.8.2). The decomposition of rock material and the rate of removal by eroding agents are determined by a combination of the tectonic conditions, the rock types, the climate history, the relief and the vegetation. If slope erosion processes work faster than the weathering processes on hillsides, then erosion is limited by the rate at which the rock is weathered. However, if the rate of weathering exceeds the rate at which transportation processes operate, deeper weathering profiles with residual soils are produced and can reach depths in excess of 100 m.

Table 3.8.1 Principal types of wet/humid tropical rain forests and their engineering implications.

| Forest type | Forest characteristics (after Prance, 2002) | Geomorphological implications for engineering |
|---|---|---|
| Cloud forest | High-altitude closed forest with many gaps, trees rarely over 2 m; often gnarled, numerous lianas, tree ferns, herbs, shrubs, epiphytes and mosses | Commonly associated with upland peat soils that take hundreds of years to develop and with little depth to unweathered bedrock; highly susceptible to instability when earthworks undertaken; sensitive to expansion of temperate crop cultivation and hill resorts; rapid erosion follows the removal of cloud forest |
| Montane rain forest | Few trees exceeding 3 m; palms and tree ferns common in undergrowth; ground layer rich in herbs and mosses, epiphytes common | Often on steep slopes with considerable amounts of surface organic matter as a result of slower decomposition in cooler conditions, but with shallow soils and shallow root systems; often in exposed areas subject to windthrow or cyclone damage; variable foundation conditions; abundant landslides |
| Lowland rain forest | Multi-layered, many trees exceeding 3 m in height, closed canopy with sparse undergrowth; frequent large lianas; relatively scarce epiphytes | Predominant forest type in areas subject to agricultural, urban and industrial development; removal leads to accelerated erosion, slope instability and increased run-off unless pre-planned protective measures are undertaken |
| Alluvial forest | Grows in seasonally inundated areas along river banks; multi-layered, closed forest with numerous gaps, buttresses and stilt roots; palms; herbaceous undergrowth; epiphytes and lianas common | Widespread in Amazonia; seasonal flooding disrupts mobility and restricts use of heavy machinery; complex soil shrink–swell and bearing problems may arise; major impacts on aquatic life |
| Swamp forest | Forest on permanently wet areas; buttresses, stilt roots and pneumatophores common; palms, ferns and herbs abundant | Often traversed by black water streams with pH values as low as 3.0 (corrosive to concrete and metals); these forests often overlie former distributary channel systems, the coarse sands and gravels of which may pose problems of stability and water retention when canals and embankments are constructed |
| Peat forest | Forest over deep peat deposits in nutrient-poor soils; rarely above 2 m in height; ground cover mostly ferns | These lowland peats are likely to shrink on drainage and develop aggressive sulphate soil conditions and moisture-sensitive clay with shrink–swell characteristics; low pH (acid) conditions may result in corrosion of structures |
| Mangrove forest | Single layered forests up to 3 m in height growing in intertidal zones in salt water throughout tropics and subtropics; evergreen stilt roots and pneumatophores common; little ground vegetation; few epiphytes | Associated with marine clays and silts, but often conceal lenses of sand derived from former river channels; greatly modified for aquaculture, port development and tourist resorts; major installations usually require deep piling unless substantial layers of stiff clay occur; potential for disruption of natural sediment fluxes and natural coastal protection causing harbour siltation or coastal erosion |

### IMPLICATIONS FOR ENGINEERING

Many humid tropical land surfaces are very old and have deep weathering profiles. However, younger and thinner regolith profiles can occur on steep and unstable slopes. Particular problems may be associated with specific rock types, such as buried karst features in limestone areas. The legacy of the past is less well understood for humid tropical areas than for most other places and planned projects therefore require careful iterative development of the site model and cautious investigation.

High mountain areas in the tropics were also subject to shifts in climate and vegetation belts during the Quaternary, when a variety of landforms and soils may have developed. High river flows, large landslides and rock-falls tend to happen more often than elsewhere in similar terrains as a result of the high volume and intensity of rainfall.

Lowland areas were affected by the changes in sea level of the Quaternary and therefore relict buried river channels and relict deposits of irregular spreads of peats, silts, sands

Table 3.8.2 Summary of Duchaufour tropical soil phases, location and climate (from Fookes, 1997a).

| Factor/conditions: Soil phase | Mineralogy | Climate needed to reach the phase | Typical locations of the phase | FAO/UNESCO equivalents (USA — Soil survey) |
|---|---|---|---|---|
| 1. Fersiallitic | Upper soils undergo decalcification and weathering of primary minerals. Quartz, alkali feldspars and muscovite not affected. Free iron usually >60% of total iron. Main clay mineral formed is 2:1 smectite; 1:1 kaolinite may appear in older well drained surfaces. With recent volcanic ashes porous andosol soils formed which are eventually replaced by 1:1 halloysites. | Mean annual temperature (°C) 13–20<br><br>Annual rainfall (m) 0.5–1.0<br><br>Dry season — Yes | Mediterranean, subtropical | Cambisols, calcisols, luvisols, alisols, andosols (alfisols, inceptisols) |
| 2. Ferruginous (ferrisols-transitional) | More strongly weathered soils form but orthoclase and muscovite typically remain unaltered.<br><br>Kaolinite is the dominant clay mineral; 2:1 minerals are subordinate and gibbsite usually absent.<br>On older land surfaces and more permeable and base rich parent material, ferrisols transitional to phase 3. Partial alteration to gibbsite may occur. | Mean annual temperature (°C) 20–25<br><br>Annual rainfall (m) 1.0–1.5<br><br>Dry season — Sometimes | Subtropical | Luvisols, nitosols, alisols, acrisols, lixisols, plintha sols (alfisols, ultisols, oxisols) |
| 3. Ferrallitic | All primary minerals except quartz are weathered by hydrolysis and much of the silica and bases removed by solution. Remaining silica combines to form kaolinite but with excess aluminium gibbsite is usually formed. Depending on the balance between iron and aluminium, iron oxide or aluminium oxide will predominate. Soils currently take 104 or more years to form. | Mean annual temperature (°C) >25<br><br>Annual rainfall (m) >1.5<br><br>Dry season — No | Tropical<br><br>Can occur in modern savanna from previous wetter climate. Conversely, some currently hot wet areas are still only in the ferruginous phase (e.g. by climate change or by rejuvenation of slopes). | |

and gravels and former coastal features are widespread. These require detailed intrusive investigation after an initial site model has been created.

## CLIMATE AND VEGETATION

The wet tropics are considered to be a challenging environment because they are particularly vulnerable to climate hazards, including floods, coastal surges, cyclones and landslides. Aggregate may be difficult to source in some localities.

The problems of engineering soils and geomorphology cannot be separated from those of the climate and vegetation (Table 3.8.1). Rain forests play a critical part in soil formation and landform stability in the hydrological cycle. Intense rainfall is a feature of the equatorial tropics. Depending on the type of forest, about 10–25% of the annual rainfall may be intercepted by foliage and returned to the atmosphere. The mean annual rainfall erosion may be 25 times greater than in humid temperate latitudes – for example, in Malaysia, the mean annual erosive power of rainfall is >25 kJ m$^{-2}$, whereas in western Britain it is just over 1.3 kJ m$^{-2}$. As elsewhere, rainfall varies with elevation and aspect. In areas prone to tropical cyclones, rainfall intensities can exceed 10 mm/h for several hours at a time and rapidly rising, long-lasting floods are common.

A tropical rain forest typically has a canopy of tall trees that rise above wetland, which becomes a swamp on level ground.

To link villages, forest dwellers use rot-proof ironwood to build raised walkways, because any heavy use turns a ground-level footpath into a mud quagmire.

Mangrove swamps thrive in salty water and muddy soils in hot, wet, coastal regions, and are distinguished by their networks of aerial roots.

Frequent flood events leave gravel and cobbles along river channels, creating strips of more stable, coarse sediments within the fine-grained soils that typify rain-forest wetlands.

An "Irish crossing", or a flat bridge, across a flood-prone river floor in low-flow conditions; it was built with no parapets so that it causes minimal encroachment into the channel and frequent floods can pass over it with no damage to its structure.

The density of equatorial rain forest precludes detailed prior ground investigation for any road or pipeline driven through it, and the necessary tree clearance turns the site into a mudbath.

Shrink-swell "black cotton" soils dominate in huge areas of flat ground that is poorly drained within the wet tropical regions.

Limestone towers, rising above plains with thin alluvial soils over limestone bedrock, characterise karst in some wet tropical regions.

Deeply weathered bedrock, and red, residual soils that thicken down-slope and into valleys, are typical of hot and wet tropical terrains.

[above] Traditional flights of terraces provide strips of flat land for cultivation and have the added benefit of curtailing soil erosion in tropical environments characterised by frequent heavy rainfall events.

[left] A modern equivalent in slope stabilisation is the planting of shrubs to hold the soil on a steep slope newly excavated on a construction site.

Red laterite and pale bauxite are residual, leached soils formed by weathering in the wet tropics.

Pinnacle karst with tall, sharp blades of limestone forms terrain that is nearly uncrossable in some hot and wet environments.

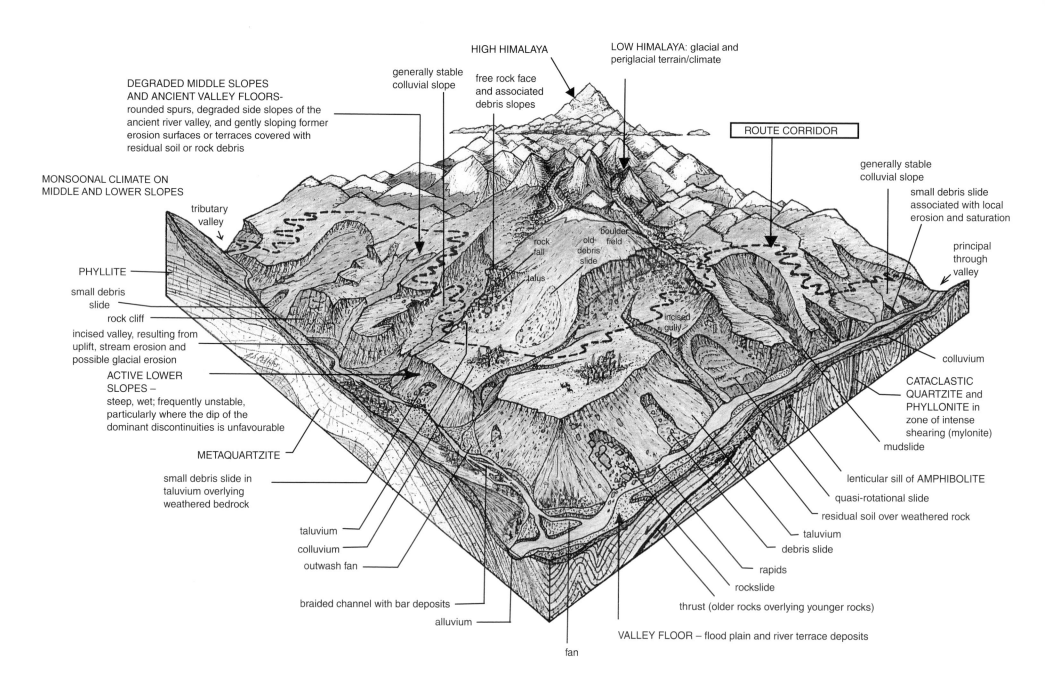

*Figure 3.9 Mountain environments.*

110

## Mountain environments (Figure 3.9)

A great variety of classification systems has been developed to capture the distinctiveness of mountain environments: the simplest are based on relative relief (e.g. hill <700 m, highland and upland 700–1000 m and mountain >1000 m). Mountains provide challenging environments for engineers, agriculturists and the development of urban infrastructure (see the photographs).

About one-third of the Earth's land surface is covered by mountains, highlands and hills. Around 10% of the world's population live in mountain regions and >40% are believed to be dependent on mountain resources, the most valuable resource being water. Mountains are extremely dynamic and sensitive environments, reflecting the combination of steep slopes, high altitudes and relative relief, together with the presence of numerous relict landforms inherited from previous advances of glaciation. This landscape sensitivity is easily disrupted by deforestation, changes in land use and construction projects, commonly leading to dramatic impacts on the scale and intensity of landslide activity and soil erosion (Charman and Lee, 2005). Large climatic variations occur depending on vertical height and horizontal location.

Many mountains have been formed in the relatively recent geological past, originating in the Tertiary Alpine orogeny; uplift is still continuing over large areas, most notably in the Karakoram Mountains. The main active mountain belts today are the Alpine–Himalayan chain (extending from Borneo through northern India into Iran, Turkey and southern Europe) and the circum-Pacific belt, including the Andes, the Rocky Mountains and the island arcs of the western Pacific. Their relative immaturity is reflected by their high relative relief, steep slopes and high rates of weathering and erosion (see Figure 1.3).

Mountain belts formed in earlier orogenic episodes (principally the Caledonian Orogeny of the Lower Palaeozoic and the Variscan, also known as the Armorican, of the Upper Palaeozoic) still exist (e.g. in Cornwall, south-western UK), but their longer history of erosion has worn them down and the extreme relative relief characteristics of the Alpine belts are no longer present. (Introduction Table 1 gives stratigraphical age names.)

### Mountain model (based on the Himalayas)

Figure 3.9 represents an active, young fold mountain and is based on cycles of high activity (driven by periods of relatively rapid uplift) that initiate periods of intense erosion as rivers cut down to lower base levels and produce steep-sided valleys. Many of these are very unstable, with the immature weathered surfaces continually being eroded. Hillslopes are typically mantled with colluvium and/or taluvium which is unstable when undercut. Screes form at the base of steep cliffs. In the intervening periods of low activity (with relatively slow uplift), continuing landslide events eventually produce shallower, more stable slopes. As erosion is less in these periods, a weathered mantle of residual soil develops to produce friable and easily erodible soils to await the next uplift. Repeated events may create a stepped landscape.

In this dynamic environment, rural management programmes or new engineering projects (e.g. roads or hill irrigation canals) require careful evaluation of landslide or erosion hazards, which are particularly likely during periods of intense rainfall (see Figure 1.3 and Table 1.3.1).

### Common engineering problems

Many of the hazards faced in mountain areas are complex, with earthquakes, snowmelt and heavy rainfall providing triggers for landslides, debris flows and flooding, which can cause widespread damage (see Table 3.9.1).

- Access and routing in steep dissected terrains present extreme difficulties in providing safe and maintained through routes and local access to villages, construction sites, and locations of borrow areas or spoil disposal. Fookes and Marsh (1981a and b), Fookes *et al.* (1985) and Hearn (2011) provide useful guides to design, construction and maintenance in the humid tropics and subtropics. The most efficient routing approach to minimize risk involves adopting 'ridge and spine' or 'ridge and spur' alignments wherever possible to avoid landslides. Where it is not possible to use spurs or ridges, it is desirable for the route to climb or descend steep slopes by tight hairpin stacks within a narrow zone that lies normal to the contours and, where possible, by selecting horizontal and vertical alignments that minimize the amounts of cut and fill. These measures minimize the exposure to landslide hazards (Figure 3.9).

- Catastrophic landslide events in mountain regions are often large and always destructive, with rapid initial rock displacement and very long run-out of debris (up to many tens of kilometres). The 1970 Huascaran disaster, for example, destroyed the town of Yungay, Peru, and killed at least 20,000 people. Such landslides are of low probability, but high consequence. Hazard assessment and avoidance of landslides are therefore the most effective means of managing risk, together with monitoring vulnerable slopes with links to early warning systems.

- Debris flows and torrents consisting of a slurry of fine and coarse material mixed with varying amounts of water are common. The mixture moves down-slope, typically along pre-existing drainage paths, in surges induced by gravity and channel bank collapse. They are especially common and highly destructive features when earthquakes, heavy rainfall or snowmelt mobilize new surface debris under a thin soil cover. Observed velocities are in the range 0.5–20 m/s.

- Natural variations related to marked diurnal fluctuations in discharge are especially common in mountain streams, reflecting both day/night changes in snowmelt and interruptions in melting conditions during periods of cloud cover or snowfall. Rain-triggered flash floods are also common, especially during the latter part of the snowmelt season. Hazards include landslide dams (e.g. from large rock avalanches), which subsequently fail or are over-topped and then rapidly eroded, in both instances releasing major floods from the impounded lake, possibly long after the landslide event. Floods can also emanate from glacial lake outbursts and snow avalanches and are a widespread winter hazard (Charman and Lee, 2005).

Construction materials are generally available from the metamorphic rocks that dominate many mountain areas. Good quarry sites are usually difficult to find. Such rocks typically show considerable variation in strength and durability. Natural rock fragments and crushed aggregates may have flaky and elongate shapes reflecting their metamorphic fabric. This factor may limit the quality of stone masonry, gabion and rock-fill, resulting in difficulties for fill compaction and in concrete and road aggregates. Hillside and alluvial particles in these areas commonly reflect the flaky nature of metamorphic rocks and usually produce poor coarse concrete aggregate and roadstone. Mica in sand used as aggregate may reduce the strength of concrete.

Important quarries, roadwork cuts, tunnels and other engineering works should always have boreholes in the GI phase of the investigation and possibly earlier and/or later as required. Good core recovery is usually difficult if not impossible in tectonized (shattered) rock common in mountains. Table 3.9.2 (after Norbury, 2010) gives practical guidance. Ground model making is difficult.

Table 3.9.1 Main factors controlling the stability of rock and soil slopes (after Hearn, 2011)

| Rock slopes | | Soil slopes | |
|---|---|---|---|
| *Conditioning factors* | *Triggering factors* | *Conditioning factors* | *Triggering factors* |
| Slope angle and height | Toe erosion by streams and rivers removing lateral support, or vertical support if undercut | Slope angle and height | Prolonged/heavy rainfall leading to a rise in groundwater level and reduction in strength |
| Rock structure orientation, including discontinuity patterns, in relation to topography (slope direction and angle – kinematic feasibility) | When degree of weathering, particularly along discontinuities, reaches a critical level (strength) | Soil depth and the presence of any adversely orientated relict structures that are derived from the original rock fabric (if in situ weathered soil) or previous failure surfaces (if taluvium/colluvium) | Intense (usually short-term) rainfall leading to saturation of surface soil layers and reduction in strength |
| Rock mass strength and weathering grade and rate of weathering | Earthquake acceleration, leading to increased driving forces | Presence of a distinct soil layer/rockhead boundary along which failure takes place | Toe erosion by streams and rivers removing lateral support |
| Presence of weak horizons within the rock mass, either more closely jointed or weaker (more clayey) layers | Heavy and/or prolonged rainfall. Increased water pressure along discontinuities | Soil composition and strength, a function of grain size, particle arrangement and mineralogy, density and moisture content | Earthquake acceleration, leading to increased driving forces |
| Presence of rock horizons/layers of varying permeability creating perched water tables | External influences including excavations, fills and spoil dumps, drainage changes | Presence of weak horizons and permanent groundwater seepages | Deforestation and other land use changes can lead to increased surface water runoff, erosion and slope instability |
| | | | External influences, including excavations, fills and spoil dumps, drainage changes |

Table 3.9.2 Scheme for recording low core recoveries. Note: Examples from Britain (after Norbury, 2010).

| Indicative core recovery | Suggested approach | Descriptive format |
|---|---|---|
| 75 – 100% | Record TCR, SCR, RQD and If as defined<br><br>Assess zones of core loss (AZCL) and assign depth ranges | Carry out full description in accordance with standards.<br><br>If core loss exceeds say 10% include statement such as 'Core loss presumed to be more weathered materials' or '...weaker materials...' or '...mudstone layers ...' as appropriate |
| 50 – 75% | Record TCR, SCR, RQD and If as defined, i.e. percentages of full run length, not recovery<br><br>Assess zones of core loss (AZCL) and assign depth ranges if possible. Where this is not possible, emphasise loss and uncertainty as to in situ conditions. | Suggested descriptive wording:<br><br>'Recovery is of stronger materials, weaker materials not recovered.'<br><br>'Recovered material is extremely weak low density white CHALK. Occasional flints.<br><br>(Structured SEAFORD CHALK)' |
| 25 – 50% | Record TCR, SCR and RQD as defined, i.e. percentages of full run length, not recovery.<br><br>Record where sensibly possible<br><br>The identification of the depth of any AZCL is unlikely to be possible | Suggested descriptive wording:<br><br>'Partial recovery. Core loss presumed to be more weathered material.'<br><br>'Recovered core comprises medium strong grey coarse grained SANDSTONE. Sand and gravel size fragments recovered. Heavy discolouration on discontinuity surfaces, penetrating up to 3 mm. (Probably Weathered COAL MEASURES SANDSTONE)' |
| < 25% | Record TCR<br><br>Leave SCR, RQD, If columns all blank. Report lengths of core sticks recovered<br><br>The identification of the depth of any AZCL is almost certainly not possible | Suggested descriptive wording:<br><br>'Minimal recovery. Core loss presumed to be more weathered material.<br><br>Recovered core comprises GRAVEL and COBBLE size fragments of strong red coarse grained granite.<br><br>(Possibly weathered PETERHEAD GRANITE)' |

Key:

TCR – total core recovery

SCR – solid core recovery

RQD – rock quality designation

If – fracture spacing or index

[left] Mountain terrain with houses and terraced fields on a spur amid slopes too steep for any development.

[above centre] Road built with hairpins up a steep slope to keep it on a spur between two landslide-prone gullies.

[above right] 'Ridge-and-spine' road that stays above unstable slopes and has attracted subsequent development.

[below left] Road temporaily blocked by a rockfall of hard, but fractured, metamorphic rock from a mountain spur.

[below centre] Cut into solid rock along a gorge wall, a road is very stable but is prone to small-scale stone-fall.

[below right] A road that suffers frequent closure due to inevitable debris falls in an active mountain environment.

Montane basin with thick sediments deposited prior to repeated tectonic uplift, rejuvenation and river incision.

[above left] Large failure of deeply weathered rock with a debris slide that undermined a hillside road.

[above centre] Mountain road cut into stable bedrock beneath a veneer of talluvium that hides any outcrop.

[right] A debris flow that reached the valley floor far below its source as a small hillside failure.

[far right] Scar of rockslide that failed on stress-relief fractures and landed in the fiord, where it created a tsunami that destroyed part of the foreground village.

[left] Debris slide from a steep forested slope that buried a road just beside its rock shelter, which was built to protect the site where the landslide hazard had previously been considered to be the greatest.

[below centre] Engineered channel through a town, designed to carry frequent floods from the adjacent mountains; normal flow is just in the central gully.

[below] Debris from this large rockslide, which was triggered by an earthquake, formed a barrier in the valley, thereby creating the foreground lake.

A terrace of alluvium and lake sediments along a valley floor upstream of a constriction formed by an ancient landslide.

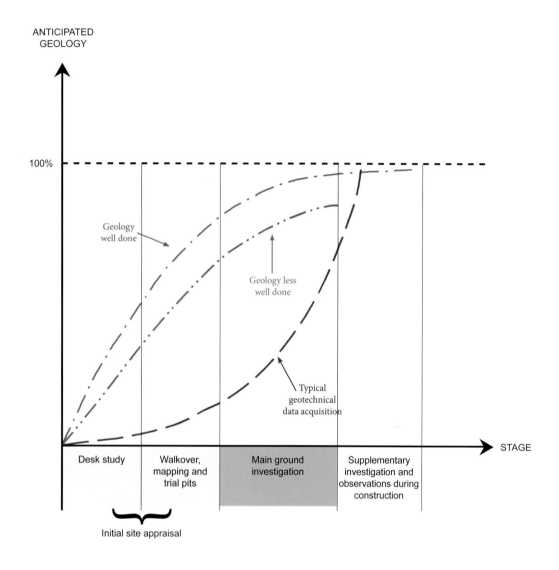

ANTICIPATED GEOLOGY

100%

Geology well done

Geology less well done

Typical geotechnical data acquisition

STAGE

Desk study | Walkover, mapping and trial pits | Main ground investigation | Supplementary investigation and observations during construction

Initial site appraisal

*Figure 4.1 Increase in site knowledge during the basic stages of a ground investigation.*

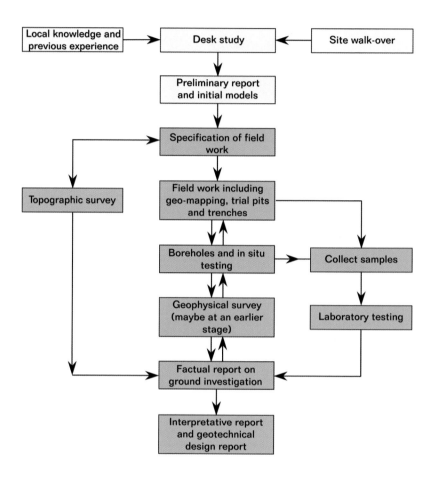

Local knowledge and previous experience → Desk study ← Site walk-over

Preliminary report and initial models

Specification of field work

Topographic survey

Field work including geo-mapping, trial pits and trenches

Boreholes and in situ testing → Collect samples

Geophysical survey (maybe at an earlier stage) → Laboratory testing

Factual report on ground investigation

Interpretative report and geotechnical design report

Table 4.1.1 Example of a simple flow chart for a basic ground investigation. Large or complex projects will have several stages of investigation.

# Part 4. Ground investigations

## Increase in site knowledge during the different stages of the ground investigation (Figure 4.1)

The principle of any investigation of the ground has to be that it is continued until the conditions are known and understood well enough for the civil engineering work to proceed safely and economically with a minimal risk of nasty surprises. A possible alternative to this, the 'observational method', is outlined in Figure 4.6. Field staff must always communicate with designers and vice versa – this cannot be emphasized too strongly.

Part 4 is devoted entirely to evaluating the increase in knowledge (information) gained about the site while investigating the site conditions in a staged (phased) manner as seen through the eyes of site models. It is not the intention here to elaborate on the field and laboratory techniques of investigation because much has been written with authority on these in Codes of Practice and Standards, and also in the many textbooks that put flesh on the bare bones of these codes and standards (e.g. Clayton *et al.*, 1995; Simons *et al.*, 2002; Norbury, 2010).

The progress of on-land ground investigations (the field and laboratory part of the broader term 'site investigation') has been the subject of numerous flow charts. In Part 4 we do not elaborate on the many activities and ramifications of flow charts (see Table 4.1.1), but instead use five basic stages of comprehensive investigations. The first stage is a *desk study*; this is followed by a *walkover survey*, preceded or perhaps followed by a *geophysical survey*, although this is not always carried out. The next stage is the *main ground investigation*, which involves intrusive investigations, such as boreholes and pits, to ascertain or confirm unseen subsurface conditions. The ground investigation may be followed by *supplementary investigations* if the previous stages have revealed specific problems.

Each site requires its own dedicated investigation, designed for that specific site. Our view is that this starts with a basic terrain model (see Figure 4.2). This becomes the initial site geomodel on which the investigation is then based, together with a knowledge of the structure to be built. As information on the site increases, improvements are made to the site geomodel, which lead to changes, as necessary, in the conduct of the site investigation. The initial terrain model therefore develops to become a more factual, detailed geological/geomorphological site model as the investigation proceeds.

Part 4 is illustrated by block models indicating how much information is gathered at each stage of the investigation (see Figures 4.2–4.6). The aim of each stage is given with a simple approach to the gathering of information at the particular stage. It is not suggested that investigations have to produce block models similar to the figures in this part – far from it. The term 'model' in the text is used to mean any way of visually portraying the subsurface and surface information, including the well-known techniques of displaying information as geological maps and cross-sections, computer simulations and geographical information systems.

The maximum value of the geomodel is in the earlier stages of the study as it enhances the quality of the design and improves the data from the main ground investigation. The accuracy of the model improves as the investigation progresses and the decisions based on the model change to become more quantitative. Decisions affecting the costs of the investigation are probably better made at the end of each stage as the investigation must be as flexible as possible. As knowledge about the site increases, so the cost of gaining more marginal information increases. There is therefore a realistic cut-off point that has to be decided for each ground investigation.

### The Figure

Figure 4.1 is based on the experience of the authors and as such is more generic than factual. It is an indicative graphical judgement of the information accrued about the site at each of the five basic stages of the investigation. It must be empha-sized that the percentages given in the figure are approximate and will vary considerably between sites.

An experienced engineering geologist working with geomorphologists and other specialists almost anywhere in the world should be able to make an initial geomodel of the site from the desk study. In developed countries such as the UK, there may be significantly more information available at the desk stage than in remote areas elsewhere in the world. We consider that it is reasonable (for an extensively mapped country) to anticipate at least 50% of the potential geological/geomorphological conditions within the desk study (see Figure 4.2). If a walkover survey is now added, then at least 65% of the conditions should be identified. It is self-evident that this is a relatively cheap way of obtaining the initial information. Expense comes with the subsequent stages. The costs given in Table 4.2.1 are based on experience and judgement of many examples, not on rigorously costed specific investigations.

At the end of the ground investigation, with the informed use of boreholes, trenches, pits, geophysics and any other appropriate methods, a minimum of about 95% of the geology/geomorphology should be known. If it is correct that a knowledge of the geology equates with the identification of potential problems, then at least 95% of the potential problems should have been identified. It would be ideal to reach 100%, but this may not be achieved on many sites until the strata are revealed in large-scale excavations. Contracts and designs should have some built-in flexibility to allow for unforeseen circumstances, with consideration given to the use of *reference conditions* (see Figure 4.2) and/or the use of the *observational method*.

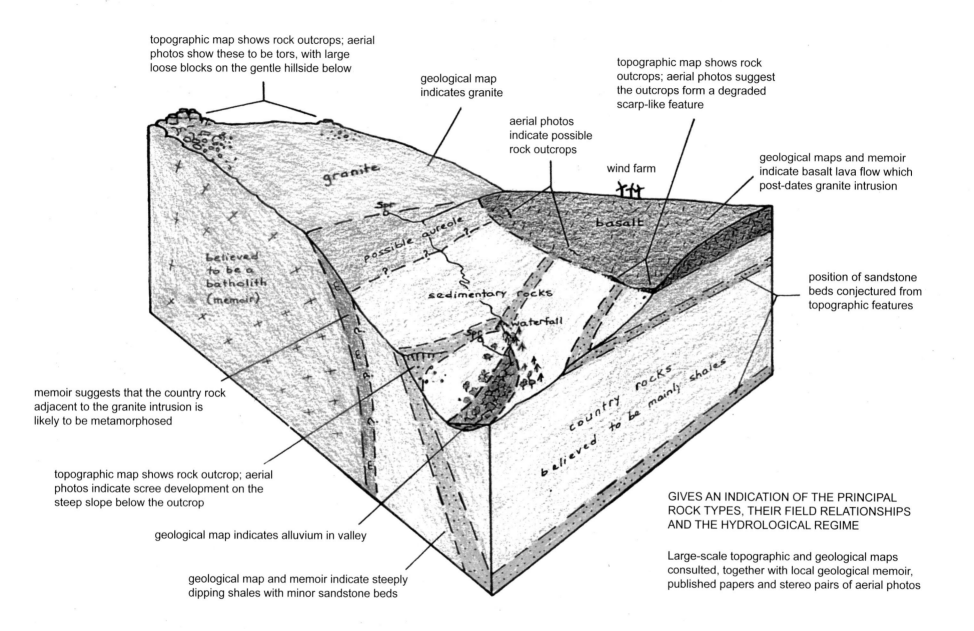

topographic map shows rock outcrops; aerial photos show these to be tors, with large loose blocks on the gentle hillside below

geological map indicates granite

topographic map shows rock outcrops; aerial photos suggest the outcrops form a degraded scarp-like feature

aerial photos indicate possible rock outcrops

wind farm

geological maps and memoir indicate basalt lava flow which post-dates granite intrusion

granite

basalt

Spr

possible aureole

sedimentary rocks

Believed to be a batholith (memoir)

position of sandstone beds conjectured from topographic features

waterfall

Spr

country rocks believed to be mainly shales

memoir suggests that the country rock adjacent to the granite intrusion is likely to be metamorphosed

topographic map shows rock outcrop; aerial photos indicate scree development on the steep slope below the outcrop

GIVES AN INDICATION OF THE PRINCIPAL ROCK TYPES, THEIR FIELD RELATIONSHIPS AND THE HYDROLOGICAL REGIME

geological map indicates alluvium in valley

geological map and memoir indicate steeply dipping shales with minor sandstone beds

Large-scale topographic and geological maps consulted, together with local geological memoir, published papers and stereo pairs of aerial photos

*Figure 4.2 Stage 1: desk study.*

## Stage 1: desk study (Figure 4.2)

The basic initial site model is conceived at the desk study stage. It is valuable for recognizing potential geohazards and designing an investigation that is both practical and reactive. The objectives of the model must be outlined at this stage, questions identified and the subsequent activities designed with these in mind. The assembled geotechnical team must be capable of defining the objectives, asking the questions and determining the ground-related activities. The following stages must be capable of delivering the answers and identifying where uncertainties remain. Knowledge of the structure to be built must play a part in the design of the site investigation procedures and techniques required.

An experienced geologist should be able to visualize, for example, a green field site, along the lines indicated in Figure 4.2, especially when geological mapping is available. For sites in areas where plenty of subsurface ground data already exist, it should be possible to picture the geology to at least a preliminary ground investigation stage.

Sources of information are listed in many publications (e.g. Perry and West, 1996; Simons *et al.*, 2002). Such information varies with the type, size and location of the site and between countries. Britain, for example, has a greater variety of historical and recorded data than most countries. Study of even the smallest site should include, at a very early stage, the examination of maps (e.g. topographical, geological, soils). If there are only very small-scale preliminary maps, or perhaps none at all, then an examination of remote sensing images and manipulation of Google Earth images are invaluable. In many parts of the world, specialist hydrological and seismic risk studies are required.

Basic sources of information include: data from governments and other authorities (e.g. published and unpublished maps, data indices, enquiry desks, libraries, published relevant refereed papers in professional and other relevant journals); local sources (e.g. residents, farmers, local societies, museums, universities, local government records and staff,

local libraries); mining records, where they exist and are confirmed to be accurate; aerial photographs, including other forms of remote sensing; and web- and computer-based information.

Table 4.2.1 is a simple guide to the estimated relative costs and benefits in each of the five basic stages of the ground investigation. Table 4.2.2 gives what Hearn (2011) considers can be learnt from 'traditional desk study sources' for mountain roads.

### TYPICAL CHANGES IN THE PERCEPTION OF GROUND CONDITIONS AS AN INVESTIGATION PROCEEDS

- *Desk study stage.* Field staff should set up a close working relationship with designers to understand the concept of the project and its engineering outline.

- *First stage of the ground investigation.* To consider all the available information relevant to the geology, geomorphology, hydrology and ground conditions of the area of interest.

- *Increase in knowledge during desk study stage.* A systematic overview of the geology, geomorphology and hydrology of the area of interest. The initial qualitative terrain model is formulated, leading to the initial geological/geomorphological site model.

- *Probable effects on the overall success of an engineering project.* This is an essential first step in any investigation of the ground conditions and allows the selection of the most suitable geological site(s) or corridor(s) for further study. The influence of the geology on the conceptual engineering design is considered.

- *Possible important increase in information obtained in geomorphologically and geologically varied and structurally complex terrains.* This gives an indication of the principal landforms, rock types, their field relationships and the hydrological regime.

*Example:* interpretation of ground conditions at this stage in karstic terrains (see Figures 5.7 and 5.8; also other figures

in Part 4). The published map indicates superficial deposits overlying limestones; previously published work suggests minor solution along discontinuities in limestone and, possibly, minor karst; at this stage judged to be only class **kI/kII** (Waltham and Fookes, 2003).

Table 4.2.1 Estimated relative costs and benefits of ground investigation.

| Approx. stage | Activity | Relative cost (C) | Information gained | Order of benefit (B) | Typical comparative B/C ratio [a] |
|---|---|---|---|---|---|
| 1 | Detailed desk study | Low | Initial knowledge of the site, avoidance of obvious problems | Very high | 2.7 |
| 2 | Site walkover | Low to medium | Visualization of the site and recognition of possible problems | Very high | 2.7–1.6 |
| 3 | Preliminary ground investigation (mainly field observation and in situ methods; can include geophysics) | Medium to high | Initial physical evaluation of the site conditions, estimation of properties, recognition and confirmation of possible problems | Depends on the accuracy of the results and can be either high or low | 2.3–0.4 [b] |
| 4 | Main ground investigation (emphasis on intrusive activities, sampling and testing) | Very high | Main quantitative information for engineering design | High to very high | 1.0 |
| 4 | Laboratory testing | Low to medium | Detailed evaluation of properties depends on ground conditions | High/medium | 2.3–1.0 [b] |
| 5 | Supplementary ground investigation | Low to medium | Increased confidence in ground conditions and particular suites of properties | High | 2.3–1.4 [b] |
| 5 | Further or special testing | Low | Increased confidence in the physical/ chemical properties tested and existing results | High | 2.3 [b] |

[a] Obtained by ascribing the following arbitrary scores to the qualitative descriptions of costs and benefits: very high, 80; high, 70; medium, 50; low, 30. These scores are then compared with the main ground investigation arbitrary score of 1.0.

[b] In many examples, the need for essential items of information would override an unfavourable B/C ratio.

Table 4.2.2 Data typically derived from traditional desk study sources (after Hearn, 2011)

| For General Engineering | | | For Landslide Identification and Assessment | | |
|---|---|---|---|---|---|
| *Topographical mapping* | *Geological mapping* | *Stereo aerial photographs* | *Topographical mapping* | *Geological mapping* | *Stereo aerial photographs* |
| Review of route corridor options in terms of topography. Identification of steep terrain. Locations of rivers and potential river crossing points. Locations of towns and villages and existing infrastructure. | Locations of major geological features (faults, shear zones, etc.). Locations of weak rocks and unstable rock structures. Potential for construction material sources to be identified (for example Fookes & Marsh, 1981). | Review of some corridor options in terms of overall topography. Identification of steep terrain. Location of rivers and potential river crossing points. Location of towns and villages and land use and existing infrastructure. | Few topographical maps show landslides, though some may show major erosion areas. Contour patterns may indicate landslide morphology. | Few published geological maps show landslide areas. Potential instability can sometimes be inferred from rock structure and rock types. Most maps indicate bedding, foliation and some joint orientations, useful for preliminary slope stability assessment. | Identification of landslides and taluvium deposits. Identification of areas of slope erosion and river scour. Tones and hues in the photography can allow wet areas to be identified, potentially relevant to landslide studies. Structural geology lineaments, bedding and major discontinuity sets may be interpreted and linked to landslide potential. Repeated aerial photography can provide information on rates of change. |

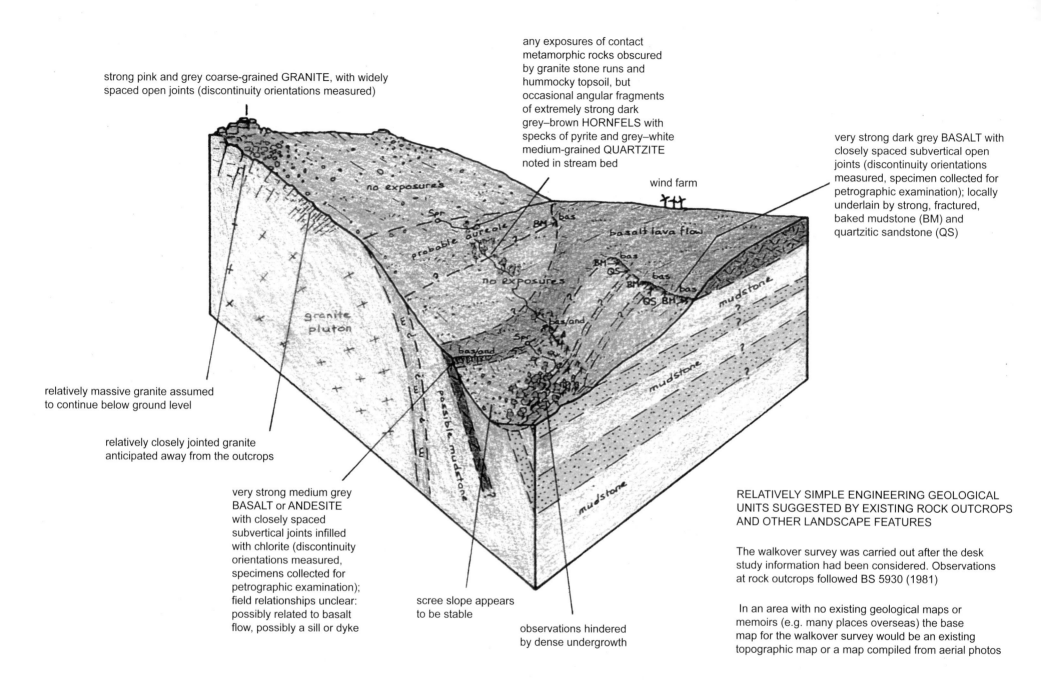

strong pink and grey coarse-grained GRANITE, with widely spaced open joints (discontinuity orientations measured)

any exposures of contact metamorphic rocks obscured by granite stone runs and hummocky topsoil, but occasional angular fragments of extremely strong dark grey–brown HORNFELS with specks of pyrite and grey–white medium-grained QUARTZITE noted in stream bed

wind farm

very strong dark grey BASALT with closely spaced subvertical open joints (discontinuity orientations measured, specimen collected for petrographic examination); locally underlain by strong, fractured, baked mudstone (BM) and quartzitic sandstone (QS)

relatively massive granite assumed to continue below ground level

relatively closely jointed granite anticipated away from the outcrops

very strong medium grey BASALT or ANDESITE with closely spaced subvertical joints infilled with chlorite (discontinuity orientations measured, specimens collected for petrographic examination); field relationships unclear: possibly related to basalt flow, possibly a sill or dyke

scree slope appears to be stable

observations hindered by dense undergrowth

RELATIVELY SIMPLE ENGINEERING GEOLOGICAL UNITS SUGGESTED BY EXISTING ROCK OUTCROPS AND OTHER LANDSCAPE FEATURES

The walkover survey was carried out after the desk study information had been considered. Observations at rock outcrops followed BS 5930 (1981)

In an area with no existing geological maps or memoirs (e.g. many places overseas) the base map for the walkover survey would be an existing topographic map or a map compiled from aerial photos

*Figure 4.3 Stage 2: walkover survey.*

## Stage 2: walkover survey (Figure 4.3)

The walkover survey is an essential part of the preliminary evaluation of the site. It should ideally be made in combination with the desk study and may be carried out by a single geologist or several members of a team, including designers, depending on the nature and size of the project. It must be well planned in advance and must include the evaluation of possible hazards; back-up plans should be made.

Fortunately, Britain is very well served with existing geological and other maps and memoirs. These are valuable in developing the initial model and even more so when they are supplemented by aerial photography, remote sensing images and the walkover survey. The walkover survey becomes more important in remoter parts of the world if there is limited availability of geological maps. In these instances, drive-over and/or fly-over surveys may be necessary (see Figures 5.1–5.4).

The initial site model at this preliminary stage is the basis on which the main ground investigation is planned to be the most cost-effective. The footprint of the engineering structure(s) should be broadly known and, if possible, there should be sufficient freedom for this to be moved to take advantage of an alternative site with more favourable ground conditions (e.g. the location of a tunnel or a large bridge to cross a valley).

Such a situation does not always develop satisfactorily when investigations are planned on the basis of a routine layout of boreholes on a grid pattern, or at set intervals on a road centreline. All too often inefficient planning of the investigation does not take account of the surface geology and fails to appreciate that the location of subsurface investigations can be used to maximize the subsurface information obtained.

A geological walkover survey should cover, as a minimum: a correlation of the ground features with existing geological, topographic and other maps; an examination of local exposures (artificial and natural); the terrain units; land use; natural physical features (e.g. escarpments, moraines, terraces or flood plains); breaks in slope; 'lumpy ground', which, in nearly all instances, is indicative of unfavourable ground conditions; distress in existing engineering or building structures; landslips; ground and surface water hydrology (e.g. sink holes, springs, solution features, seepages, stream levels, potential river flood plain erosion, avulsion); and, where relevant, the potential for seismic and volcanic activity.

### TERRAIN EVALUATION

The technique of terrain evaluation has a precise role in preliminary studies, depending on the location and circumstances. Terrain evaluation (or analysis) is a scientific interpretation of the landforms, vegetation and soils of a given area in relation to the uses to which it may be put.

This should help progress towards an initial understanding of the natural features in the landscape by dividing it into meaningful distinct homogeneous units. The recognition of such terrain units, at whatever scale is considered, implies that there is a genetic relationship between the landforms and the processes and materials involved in their development (Grant, 1968; Lawrance, 1972; Lawrance *et al.*, 1993). It is essentially the same approach as that used in the initial photo-interpretation, mapping or drive-over/walkover of any site. Its value is highest in remote locations with few or no existing maps (see also Figure 4.5).

### TYPICAL CHANGES IN THE PERCEPTION OF GROUND CONDITIONS AS THE INVESTIGATION PROCEEDS

- *Requirements of the walkover stage.* This is the second stage of the preliminary ground investigation and is used to obtain information on the engineering geology of the site and to establish whether further investigations are required. It may also include limited subsurface exploration and the acquisition of local knowledge (e.g. talking to farmers).

- *Increase in knowledge during the walkover stage.* This involves the confirmation or initial prediction of the ground conditions likely to be encountered during construction or development. Potential problems such as slope instability can be identified and any gaps in knowledge recorded (e.g.

the absence of exposures of suspected weak rocks and fault zones). Engineering geological/geomorphological units are recognized and a semi-quantitative (judgemental plus simple index tests) evaluation of the engineering properties can be made. The main ground investigation and any geophysical investigation can then be designed in detail. Construction of the geotechnical model is started.

- *Likely effects on the overall success of an engineering project.* It is necessary to obtain a feel for the site and to recognize the main engineering difficulties. This phase establishes the requirements and optimum techniques for further investigations. Significant facts about the structure of the ground and the properties of the materials may not be revealed and therefore there is a possibility of unforeseen ground conditions, construction problems and claims without further study. Conceptual designs should be improved in the light of observations made during the walkover.

- *Make special note of 'lumpy ground'.* The processes or situations that lead to the formation of lumpy ground include sink holes, old mines, old landslides, solifluction activity, moraines and other glacial features and mine or quarry waste. Such situations are likely to require a specially designed specific and focussed investigation.

- *Possible increase in mapping and imagery information in geologically varied and structurally complex terrains.* Relatively simple engineering geological/geomorphological observations can be made on existing rock outcrops and other landscape features.

*Example*: interpretation of ground conditions at this stage in karstic terrains continued from Figure 4.2 (see also Figures 5.7 and 5.8). Occasional sink holes observed; simple probe drilling indicates major local solution along discontinuities; at this stage judged to be **class kII/kIII**; changes made to conceptual engineering.

Horizontal scars and terraces identify a series of sedimentary rocks with alternating weak shales and strong limestones.

[above] At outcrop, the top metre of nearly vertical slates has been buckled where the hillside's surface zone is creeping downslope.

[left] Gently dipping chalk forms a rounded escarpment with its scarp face overlooking lowland formed on an underlying clay.

A cut face exposes completely weathered granite beneath a thick layer of residual soil that is largely ferricrete and therefore more resistant than the rock beneath.

[left] An inclined fault forms the boundary between pale, weathered volcanics and darker, fresher basalt.

[above] An active fault still forms a fresh scar 20 years after causing an earthquake when it moved 5 metres.

[above right] Tension cracks in a road are caused by a small fault reactivated during coal mining subsidence.

The entire foreground, crossed by and lying left of the road, is large-scale lumpy ground that identifies a complex of landslide forms and debris that are still moving.

"Lumpy ground" is a terrain feature that is readily identified on a walkover survey, and is generally indicative of some variety of difficult ground conditions; at this site the hollows are backfilled, open or inadequately capped shafts, and the hillocks are unconsolidated mine debris.

[below left] Old mine workings along two veins are recognisable by the lines of shadowed pits.

[below middle] A walkover survey in dense rain forest is severely limited, but is still invaluable.

Lumpy ground can be formed by sinkholes in the soil cover on limestone karst.

A walkover at this site should recognise the low, rounded mounds of glacial till on the right, the single, steeper hill that is a colliery tip-heap on the left, and the central area of flat, wet ground that is a peat bog; rockhead beneath the peat was breached by a rising mine heading, causing a catastrophic inrush of saturated soil debris.

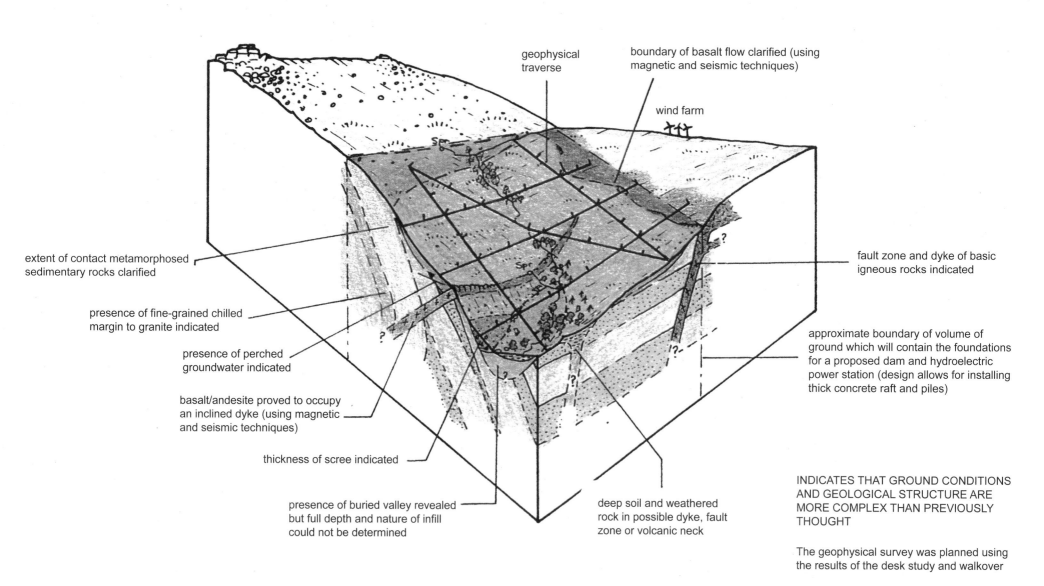

geophysical traverse

boundary of basalt flow clarified (using magnetic and seismic techniques)

wind farm

extent of contact metamorphosed sedimentary rocks clarified

presence of fine-grained chilled margin to granite indicated

presence of perched groundwater indicated

basalt/andesite proved to occupy an inclined dyke (using magnetic and seismic techniques)

thickness of scree indicated

presence of buried valley revealed but full depth and nature of infill could not be determined

deep soil and weathered rock in possible dyke, fault zone or volcanic neck

fault zone and dyke of basic igneous rocks indicated

approximate boundary of volume of ground which will contain the foundations for a proposed dam and hydroelectric power station (design allows for installing thick concrete raft and piles)

INDICATES THAT GROUND CONDITIONS AND GEOLOGICAL STRUCTURE ARE MORE COMPLEX THAN PREVIOUSLY THOUGHT

The geophysical survey was planned using the results of the desk study and walkover

*Figure 4.4 Stage 3: shallow geophysical survey.*

126

## Stage 3: shallow geophysical surveys (Figure 4.4)

Shallow geophysical techniques are constantly improving and can offer increased benefits as part of geotechnical investigations in civil engineering projects. They are not always used in ground investigations, but they should always be considered as a potential tool. Geophysics can give a valuable overall picture, but these techniques do need ground truthing i.e. confirmation on the ground in conjunction with boreholes, pits and other intrusive techniques to help calibrate the subsurface profiles or to test-drill specific identified anomalies (Bibliography, Group B books; Waltham, 2009).

Geophysical surveys have two main uses in ground investigations: (1) searching a large area for anomalies before drilling and (2) correlating the strata between boreholes.

Geophysical surveys are relatively low cost compared with multiple boreholes and can be cost-effective in site investigations of difficult ground situations where a specific type of survey might be appropriate. There is no single geophysical survey that is applicable to all problems. Geophysical techniques can now be applied effectively to particular situations – for example, using magnetic surveys to search for suspected mine shafts. All geophysical surveys require discussion with independent geophysical scientists to obtain advice before and during discussions, and with specialist geophysical ground investigation contractors to interpret the results.

Geophysical exploration techniques involve the remote sensing of some physical property of the ground, either using instruments that remain on or near the ground surface, or instruments that are inserted into boreholes during drilling. Aerial surveys can be carried out, but are usually reserved for extremely large or linear remote sites. Passive methods accurately measure the ground properties and search for small anomalies (distortions) within the overall pattern. These include gravity and magnetic surveys. Induction methods send a signal into the ground and pick it up again nearby. These include seismic, electrical, electromagnetic and radar surveys.

Modern techniques fall into broad groups and each technique can have its uses in specific ground investigations.

- *Seismic surveys* use shock waves produced by hammer blows or explosions; these are reflected or refracted at geological boundaries (essential in oil exploration) and are applicable to ground investigations as the signals relate to the strength of the rock mass or soil.

- *Magnetic surveys* record distortions of the Earth's magnetic field and are notably successful in locating old mine shafts.

- *Gravity surveys* record minute variations in the Earth's gravitational force and therefore identify low-strength porous rocks and natural or artificial ground cavities.

- *Electrical surveys* include many methods and are widely applied successfully in mineral exploration; they can cover large areas to identify anomalies or contrasts that require further investigation.

- *Electromagnetic surveys* create an electromagnetic field in the ground and measure differences in the magnetic properties of ground materials, yielding data similar to that from electrical surveys.

- *Ground-probing radar* is a special type of electro-magnetic survey that has a limited depth of penetration, but can be useful in shallow investigations.

### TYPICAL CHANGES IN THE PERCEPTION OF GROUND CONDITIONS AS THE INVESTIGATION PROCEEDS

- *Requirements of the geophysics stage.* Simple shallow geophysical techniques may be used during the preliminary investigation, or more detailed surveys may form part of the following main investigation. These are used to obtain basic information on the subsurface geology relatively easily and quickly.

- *Increase in knowledge during investigation stage.* Geophysical surveys extends our knowledge of the subsurface geology provided that the techniques used are appropriate for the terrain. The results must be interpreted carefully and must be verified by drilling and pitting. Features such as buried valleys may be identified, predictions modified, initial calculations of the amounts of materials made and the initial site geomodel updated.

- *Likely effects on the overall success of an engineering project.* Geophysical surveys reduce uncertainty in the predicted ground conditions and may therefore improve confidence in design decisions; they should be used to help optimize the location and depth of boreholes.

- *Possible increase in information obtained in geologically varied and structurally complex terrains.* Geophysical surveys can indicate whether the ground conditions and geological structures are more (or less) complex than previously thought. Changes to the conceptual design and/or further investigation can then be considered.

*Example:* interpretation of ground conditions at this stage in karstic terrains, continued from Figure 4.3 (see also Figures 5.7 and 5.8). An extensive geophysical survey of the whole site indicates the position of the water-table and suggests a locally deep overburden; however, further interpretation of the results is inconclusive. Possibly **class kIII**. Identifies a requirement to increase the density of intrusive subsurface investigation techniques.

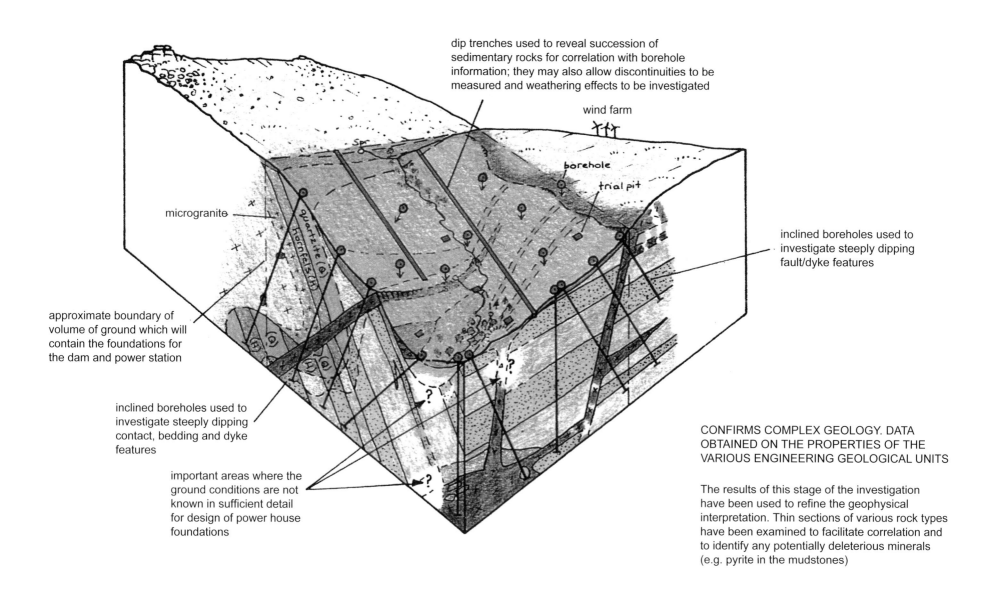

dip trenches used to reveal succession of
sedimentary rocks for correlation with borehole
information; they may also allow discontinuities to be
measured and weathering effects to be investigated

wind farm

borehole

trial pit

microgranite

inclined boreholes used to
investigate steeply dipping
fault/dyke features

approximate boundary of
volume of ground which will
contain the foundations for
the dam and power station

inclined boreholes used to
investigate steeply dipping
contact, bedding and dyke
features

important areas where the
ground conditions are not
known in sufficient detail
for design of power house
foundations

CONFIRMS COMPLEX GEOLOGY. DATA
OBTAINED ON THE PROPERTIES OF THE
VARIOUS ENGINEERING GEOLOGICAL UNITS

The results of this stage of the investigation
have been used to refine the geophysical
interpretation. Thin sections of various rock types
have been examined to facilitate correlation and
to identify any potentially deleterious minerals
(e.g. pyrite in the mudstones)

*Figure 4.5 Stage 4: main ground investigation.*

## *Stage 4: main ground investigation (Figure 4.5)*

The function of the model during the main investigation is to ensure that the geological situation is understood as completely as possible and that no major surprises will be discovered at the subsequent stages of design or construction. It is therefore necessary for a comprehensive understanding to be obtained of the geological and geomorphological processes occurring at the site and even beyond its boundaries. Geological mapping and remote sensing interpretation are key initial components of this understanding, especially for large or linear project sites.

Ideally, the objectives identified in the earlier stages should be achieved and crucial questions answered. At this stage, or preferably earlier, it should be decided whether enough information has been obtained or could safely be assumed to be obtainable. This information provides the geological and geomorphological basis for rational design, including the use of the design events and *reference conditions* (Table 4.5.1) within the contract (Baynes *et al.*, 2005). The reference conditions approach is particularly helpful for very large or linear projects and provides a template for defining unforeseen/unforeseeable conditions (Fookes, 1997a). Another approach would be to use the *observation method* (see Figure 4.6).

### GROUND INVESTIGATION TECHNIQUES

Much has been written about the many site investigation techniques used for different ground situations and the various problems that can occur. We will summarize these simply, and will also briefly explore the use of the geomodel in investigating the various features of a site.

A full geological understanding allied to the planning and layout of ground investigations often receives only scant attention. Figures 4.7 and 4.8 will help in the design of the layout and planning of investigations, but it is essential to be mindful of the model developed by this stage and the local environment (see Parts 1, 2 and 3).

Codes and standards concentrate on the instruments, instrumentation and machinery of drilling, boring and pitting.

Table 4.5.1 Function of the reference conditions, adapted to incorporate geological and geomorphological information.

| Aim: to document the range of ground conditions that can reasonably be foreseen for contract purposes, especially design and payment | |
| --- | --- |
| 1 | Formally define and describe the components of the site model (including surface processes, materials and landforms). This will most likely involve erecting and defining geological and geomorphological reference conditions within the contract by grouping together terrain units (Figure 4.3) with similar engineering characteristics |
| 2 | Simplify the geomorphological processes into a series of events that define the basis for design (e.g. the 1 in 100-year flood; a landslide event of a particular size and intensity) |
| *Use of reference conditions can:* | |
| 1 | Allow a reduction in the overall laboratory testing schedule as only representative and extreme samples from each reference condition have to be tested, rather than testing all the terrain units encountered |
| 2 | Allow the incorporation of knowledge from similar terrain units that occur outside the project area to be usefully correlated with the reference conditions |
| 3 | Be of practical help during construction in anticipating ground conditions and predicting equipment performance and capability |

The typical equipment in current use is generated by the need for soil or rock penetration, core recovery and sampling. The need for high-quality observations (e.g. the logging of cores, pits and samples) remains paramount, in addition to the skilled use of site instrumentation such as piezometry and measurements of settlement, strain and pressure. Developments in investigation techniques over the last three decades include cone penetration and other systems of determining soil and rock properties in situ. Good textbooks (e.g. Clayton *et al.*, 1995; Simons *et al.*, 2002; Norbury, 2010) are fairly exhaustive on these subjects, but can never be completely up to date. The trilogy of extremely comprehensive manuals on soil laboratory testing (Head, 2006; Head and Epps, 2011, 2014) should be a first reference for any laboratory testing based on British Standards.

The ability to conceive geomodels to illustrate various points concerned with the relatively large-scale geometry of the overall geology and geomorphology is effectively endless. It is worth saying again (to emphasize what every good ground investigator knows) that each site deserves its own specifically designed investigation, dependent on the proposed engineering structures and the anticipated geological conditions of the site. Larger and more complex sites may require specialist input e.g. Hearn (2011) shows in Table 4.5.2 what he considers may be needed for a mountain road.

The geomodel is also particularly useful in establishing the likely presence of small-scale subsurface geological features (see Figures 4.7 and 4.8, and Part 5), which may have a geometrical relationship with the overall geology and may be crucial in the design – for example, the orientation of slick discontinuities. Figures 1.8 and 1.9 give pictorial block models of such geological features.

### TYPICAL CHANGES IN THE PERCEPTION OF GROUND CONDITIONS AS THE INVESTIGATION PROCEEDS

- *Requirements of the ground investigation stage.* Intrusive techniques (e.g. boring, drilling and digging) are used along with in situ and laboratory testing to define the ground conditions and to obtain accurate data with which to calculate the design parameters.

- *Increase in ground knowledge during the investigation stage.* This stage further extends and refines the engineering data generated for the design. The results of earlier stages are reconsidered and the characteristics of the engineering geological/geomorphological units are updated. The engineering properties are quantified and detailed calculations are made of the amounts of materials; the site geomodel and geotechnical model are thus improved. The investigation is modified as an increase in the knowledge of the ground is obtained.

- *Probable effects on the overall success of an engineering project.* The selection of the most appropriate techniques and high standards of logging, testing and interpretation will contribute to the selection of reliable design parameters and enable cost and time schedules to be estimated with reasonable accuracy.

- *Possible increase in information obtained in geologically varied and structurally complex terrains.* This stage will confirm the complex (or otherwise) subsurface geology. Data obtained on the properties of the various engineering geological/geomorphological units may make changes to earlier engineering designs necessary (see Figures 5.1 and 5.2).

*Example:* interpretation of ground conditions at this stage in karstic terrains continued from Figure 4.4 (see also Figures 5.7 and 5.8). Drilling confirms the presence of infilled dolines and locates several cavities above and below the water-table: **class kIII/kIV**. This represents a significant worsening of the sub-surface conditions envisaged in the early stages of the model production.

Table 4.5.2 Common specialist skills for the assessment of terrain and slope stability and the design of mountain roads (Hearn, 2011).

| Specialist | Terrain classification | Landslide mapping | | Identifying areas of future instability | Ground investigation | | | Slope stability assessment and analysis | | Design of engineering works | | | | | | |
|---|---|---|---|---|---|---|---|---|---|---|---|---|---|---|---|---|
| | | From remote sensing | From field observation | From remote sensing and field observation | Planning | Supervising | Interpreting | Soil slopes | Rock slopes | Alignment | Earthworks | Soil slope stabilisation | Rock slope stabilisation & protection | Retaining walls | Drainage | Erosion protection |
| Geologist | likely | main | main | main | main | main | main | unlikely | main | main | main | unlikely | unlikely | unlikely | unlikely | unlikely |
| Geomorphologist | main | main | main | main | likely | main | main | main | main | likely | unlikely | unlikely | unlikely | unlikely | main | unlikely |
| Engineering geologist | likely | likely | main | main | main | main | main | main | main | likely | main | likely | main | likely | main | main |
| Geotechnical engineer | unlikely | unlikely | main | main | likely | likely | main | main | main | main | main | main | main | main | likely | main |
| Civil engineer (roads and structures) | unlikely | unlikely | unlikely | unlikely | unlikely | unlikely | unlikely | unlikely | unlikely | main | likely | likely | likely | likely | likely | likely |
| Drainage engineer | unlikely | unlikely | unlikely | unlikely | unlikely | unlikely | unlikely | unlikely | unlikely | unlikely | unlikely | unlikely | unlikely | unlikely | main | main |
| Bioengineer/forester | unlikely | unlikely | main | unlikely | unlikely | unlikely | unlikely | likely | unlikely | unlikely | unlikely | unlikely | unlikely | unlikely | likely | main |

| | |
|---|---|
| ■ | main skill fields |
| ▫ | some skills likely |
| ▨ | some skills possibly |
| □ | skills unlikely |

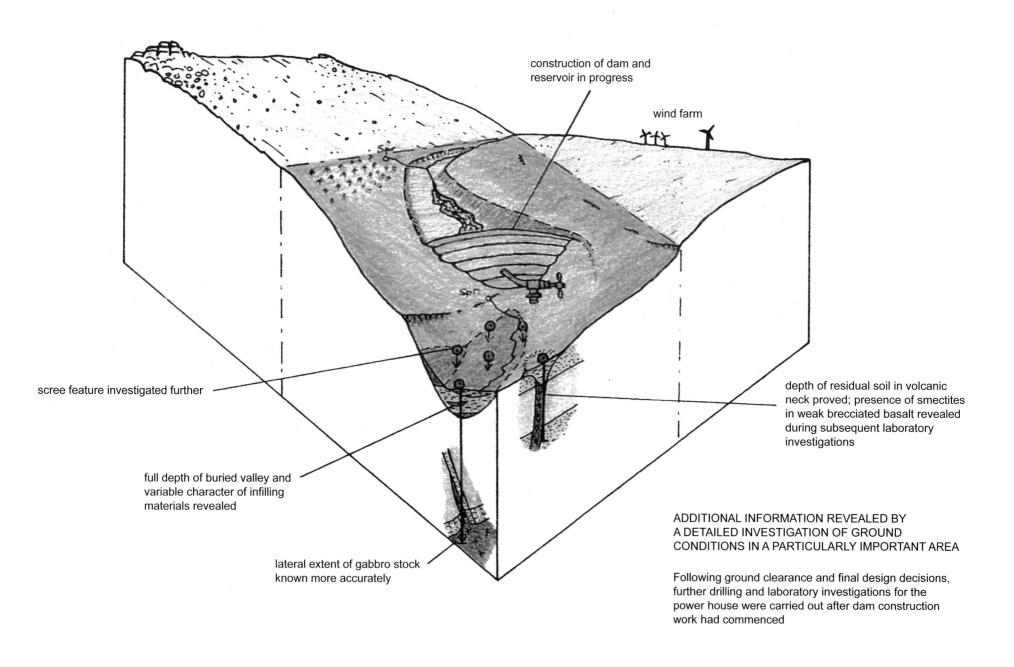

construction of dam and
reservoir in progress

wind farm

scree feature investigated further

depth of residual soil in volcanic
neck proved; presence of smectites
in weak brecciated basalt revealed
during subsequent laboratory
investigations

full depth of buried valley and
variable character of infilling
materials revealed

ADDITIONAL INFORMATION REVEALED BY
A DETAILED INVESTIGATION OF GROUND
CONDITIONS IN A PARTICULARLY IMPORTANT AREA

Following ground clearance and final design decisions,
further drilling and laboratory investigations for the
power house were carried out after dam construction
work had commenced

lateral extent of gabbro stock
known more accurately

*Figure 4.6 Stage 5: additional ground investigations.*

## Stage 5: additional ground investigations (Figure 4.6)

Additional ground investigations are not always carried out. The site may be sufficiently understood not to require further investigation, or it may not be possible to remobilize a drilling rig, or there may be compelling cost or contractual reasons not to do so. Contracts for major projects can be written so that extensions to the main ground investigation can easily be carried out if required, or the main investigation may be designed to be modified and extended as more subsurface information is obtained during the early stages of the ground investigation. Examples of complex ground are illustrated in Figures 4.8, 5.2, 5.6 and 5.9.

The layout of subsurface investigation techniques is described further in Figures 4.7 and 4.8. These have been drawn to illustrate the relation between subsurface techniques and the spatial distribution of geological features within the model.

The aim of the completed ground investigation, as always, is to maximize the collection of data and to be cost-effective. The density of boreholes and pits could be increased if the findings indicate that further investigation is necessary. In variable ground, a balance has to be struck between the cost of additional investigations and the value to the designer and constructor of the information that it is likely to yield. The more variable the ground, the less useful information might be obtained from any one borehole or test pit and therefore there will not necessarily be an improvement in the ability to extrapolate strata from adjacent boreholes or pits. Geophysical techniques improve extrapolation and the geological model will certainly help in achieving this.

Alternatively, it may be decided that an observational method (Table 4.6.1) would be the most appropriate way of solving potential design and construction problems based on inadequate geotechnical information, or because further, but better, experience and/or time-consuming investigation has little chance of improving the information available. It is

Table 4.6.1 Sequence of important features of the observational method to be used during construction (after Nicholson *et al.*, 1999; Baynes *et al.*, 2005).

| | |
|---|---|
| 1 | Exploration sufficient to establish at least the general nature, pattern and properties of the deposits, but not necessarily the detail |
| 2 | Assessment of the most probable conditions and the most unfavourable conceivable deviations from these conditions; in this assessment, geology often plays a major part |
| 3 | Establishment of the design based on a working hypothesis of behaviour anticipated under the most probable conditions |
| 4 | Selection of quantities to be observed as construction proceeds and calculation of their anticipated values on the basis of the working hypothesis |
| 5 | Calculation of values of the same quantities under the most unfavourable conditions compatible with the available data concerning the subsurface conditions |
| 6 | Selection in advance of a course of action or modification of design for every foreseeable significant deviation of the observational findings from those predicted on the basis of the working hypothesis |
| 7 | Measurement of quantities to be observed and an ongoing evaluation of the actual conditions |
| 8 | Modification of design to suit the actual conditions |

important to monitor the stability of earthworks, particularly cut slopes, at regular intervals following completion of the works.

Concern is often expressed about delays and the escalating costs of construction, not only in Britain, but also in other parts of the world, and suggestions, however justified, of additional investigations may be frowned upon. Delays are often attributed to inadequate site and ground investigation. Simons *et al.* (2002) attribute the primary causes of these shortcomings in the ground investigation to include the following: unfair or unsuitable methods of competition; inappropriate conditions of the contract; insufficient and inadequate supervision; inadequate and unenforceable specification of work; lack of client awareness; inadequate finance; insufficient time to carry out a proper investigation; and a lack of geotechnical expertise.

### TYPICAL CHANGES IN THE PERCEPTION OF GROUND CONDITIONS AS THE INVESTIGATION PROCEEDS

- *Requirements of the additional investigation stage.* This stage aims to remove the uncertainty resulting from incomplete or suspect data, to confirm the proposed course of action, or to support design changes. It will also warn of special studies that may be required during the construction or development of the site – for example, intensive targeted drilling, a study of excavations, full-scale field trials or laboratory investigations.

- *Increase in knowledge during the additional investigation stage.* This stage completes the site model, perhaps with one or more additional specialized geological/geomorphological models of all or a particular part of the site. It monitors the performance of slopes, foundations and structures and reveals more precisely the amounts of materials and the accuracy of the predictions of the design parameters, in addition to improving the geotechnical model.

- *Probable effects on the overall success of an engineering project.* Redesigning structures or foundations may increase the contract time and costs, but may also increase the service life and reduce long-term remedial and maintenance costs.

- *Possible increase in information obtained in geologically varied and structurally complex terrains.* Additional, possibly unexpected, information may be revealed by a detailed investigation of the ground conditions in a particularly important area (see Figures 5.1 and 5.2).

*Example:* interpretation of ground conditions at this stage in karstic terrains, continued from Figure 4.4 (see also Figures 5.7 and 5.8). Major development of dolines and extensive cave system revealed in an area of particular interest: now **class kIV**.

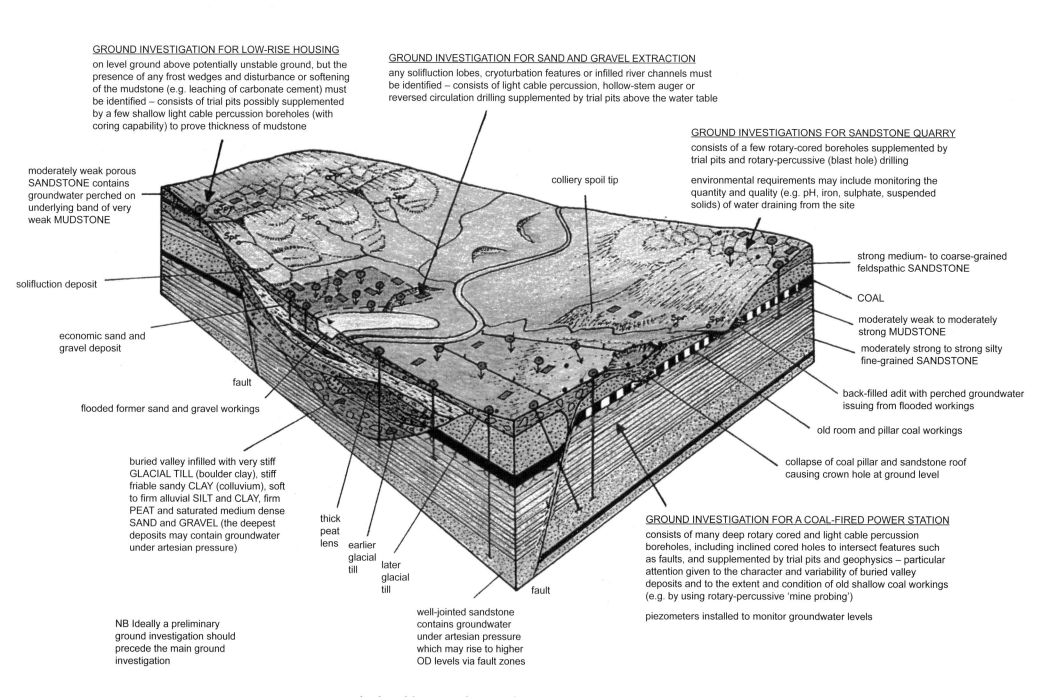

GROUND INVESTIGATION FOR LOW-RISE HOUSING

on level ground above potentially unstable ground, but the presence of any frost wedges and disturbance or softening of the mudstone (e.g. leaching of carbonate cement) must be identified – consists of trial pits possibly supplemented by a few shallow light cable percussion boreholes (with coring capability) to prove thickness of mudstone

GROUND INVESTIGATION FOR SAND AND GRAVEL EXTRACTION

any solifluction lobes, cryoturbation features or infilled river channels must be identified – consists of light cable percussion, hollow-stem auger or reversed circulation drilling supplemented by trial pits above the water table

GROUND INVESTIGATIONS FOR SANDSTONE QUARRY

consists of a few rotary-cored boreholes supplemented by trial pits and rotary-percussive (blast hole) drilling

environmental requirements may include monitoring the quantity and quality (e.g. pH, iron, sulphate, suspended solids) of water draining from the site

moderately weak porous SANDSTONE contains groundwater perched on underlying band of very weak MUDSTONE

colliery spoil tip

strong medium- to coarse-grained feldspathic SANDSTONE

COAL

solifluction deposit

moderately weak to moderately strong MUDSTONE

moderately strong to strong silty fine-grained SANDSTONE

economic sand and gravel deposit

back-filled adit with perched groundwater issuing from flooded workings

fault

old room and pillar coal workings

flooded former sand and gravel workings

collapse of coal pillar and sandstone roof causing crown hole at ground level

buried valley infilled with very stiff GLACIAL TILL (boulder clay), stiff friable sandy CLAY (colluvium), soft to firm alluvial SILT and CLAY, firm PEAT and saturated medium dense SAND and GRAVEL (the deepest deposits may contain groundwater under artesian pressure)

thick peat lens

earlier glacial till

later glacial till

GROUND INVESTIGATION FOR A COAL-FIRED POWER STATION

consists of many deep rotary cored and light cable percussion boreholes, including inclined cored holes to intersect features such as faults, and supplemented by trial pits and geophysics – particular attention given to the character and variability of buried valley deposits and to the extent and condition of old shallow coal workings (e.g. by using rotary-percussive 'mine probing')

piezometers installed to monitor groundwater levels

NB Ideally a preliminary ground investigation should precede the main ground investigation

well-jointed sandstone contains groundwater under artesian pressure which may rise to higher OD levels via fault zones

fault

*Figure 4.7 Idealized layout of ground investigations in gently dipping strata.*

## Idealized layout of ground investigations in gently dipping strata (Figure 4.7)

Figures 4.7 and 4.8 are simple examples of composite pictorial block models showing fairly normal subsurface situations which, although generally applicable, could be from Britain or many other temperate environments (see Figures 1.1, 1.2, 3.3 and 3.4). It is emphasized that these are only hypothetical basic models and many such models of different geological associations could be quickly sketched. The examples have been annotated to show conceptual ground investigations for various engineering projects.

### The Figure

Figure 4.7 shows a stratified Coal Measure bedrock sequence of sandstones, shales and other sedimentary rocks, intersected by ancient faults and incised by a deep Quaternary buried valley infilled with glacial and fluvial deposits.

The following general points can be made about Figure 4.7. However, any number of specific details could be developed.

- After the desk study and walkover survey, a small number of widely spaced boreholes of approximately equal spacing could be used to give the basic stratigraphic sequence and weathering profile of the bedrock. A geophysical survey should be considered (in this instance, a seismic survey) to determine the depth of the bedrock using the boreholes to aid interpretation. The seismic survey should be carried out on a grid plan, or at least between the borehole locations. Additional boreholes could be sunk in areas of geophysical anomalies.

- The use of exploratory pits and trenches is a relatively cheap and quick method of retrieving information from shallow depths; at these depths, they may be more effective than boreholes. Success depends on the type of material, the groundwater conditions, the depth of the bedrock and the variability of the superficial deposits. Costs increase for pits or trenches >1.5 m deep because of the need for internal support (which may inhibit logging) and extra supervisory staff.

Deeper unsupported trial pits are used solely to obtain composite bulk samples of potential construction materials above the water-table and not for close inspection of the in situ strata. They are very cost-effective.

- The topography should be considered. For example, exploratory trenches or pits dug parallel in a downhill sequence not only reveal the lateral and vertical variability of the superficial deposits, but also indicate in detail the vertical variability and weathering of the underlying bedrock.

- A sequence of trenches, dug parallel to one another and normal to the direction of any suspected shallow buried valleys, should be suitable for locating these valleys as well as for sampling the ground conditions. The spacing of the trenches is dependent on the site and any success in locating buried valleys depends on the thickness of the valley-fill (i.e. colluvium or terrace alluvium). Boreholes should replace trenches where the alluvium is thick. If the likely position of a buried valley is not known, a geophysical survey should be carried out in a grid pattern with a relatively wide and equal spacing of boreholes. These could then highlight the areas that require additional investigation. Buried valleys should always be considered possible in any projects close to the sea and in glacial terrains. Sequences of trenches/boreholes are also worth considering for locating shallow former mine sites, old waste dumps and back-filled mine shafts.

Figure 4.7 shows that vertical boreholes sunk in relatively flat-lying beds reveal the stratigraphy of the bedrock. The depth and spacing are dependent on the site conditions and the type of engineering project, as indicated in the specific projects annotated on the model. As a general rule, in flat-lying strata the boreholes can be fairly widely spaced. They should be drilled to a depth below that of the engineering foundations. This allows the stratigraphic horizons to be correlated between boreholes and exploration of the groundwater conditions below the foundations.

Coal Measures in an opencast mine, with horizontal, dark grey mudstones and buff sandstones above a coal seam at floor level; this method of mining is economical where overburden is less than about 20 times the seam thickness.

[below left] A truck-mounted hydraulic drill at a ground investigation, coring bedrock beneath a few metres of cased hole through overburden.

[upper right] A back-filled mine shaft 3 metres in diameter, recognisable by the darker fill where colliery waste about a metre deep had been stripped from the site, after the shaft was located by trenching through the cover across the area where it was suspected to lie.

[middle right] The end of a house that collapsed into an old mine shaft when its backfill had run into the deeper workings; when the houses were built, the filled shaft had unfortunately been obscured by colliery waste during site restoration.

[lower right] Part of the store of old mine plans held by the UK Coal Authority Mining Records Office; old shafts are a major hazard in the coalfields, and though the data are known to be incomplete, a search of them is the best first stage in any investigation of mined ground.

[right] Rockhead exposed in the bank of a small river, with gently dipping dark grey shales beneath just a few metres of colluvium and poorly sorted alluvial sediments.

[right] Exposed in a cut face along the alignment of a new road, a minor fault in horizontal Coal Measure rocks displaces the buff and rusty sandstones, the dark grey shales and the single coal seam that lies at floor level beyond the fault.

[above left] Microgravity survey searching for negative anomalies that indicate the presence of mined cavities at shallow depth in sandstone.

[above middle left] Probing with a hand-held drill can reach 6 metres deep in searches for mined or natural cavities.

[above middle right] Rotary coring to depths of 50 metres in strong limestone to identify ground conditions along the line of a proposed tunnel.

[above far right] Drill core is logged on site, and is then boxed for later recording and sampling.

[near right] A trial pit provides the best information on the top few metres of the soil profile, but a pit requires support struts if it is taken any deeper.

[far right] The problem with a trial pit in rain forest is that it is likely to fill with water, even before the deeper parts of its walls can be inspected.

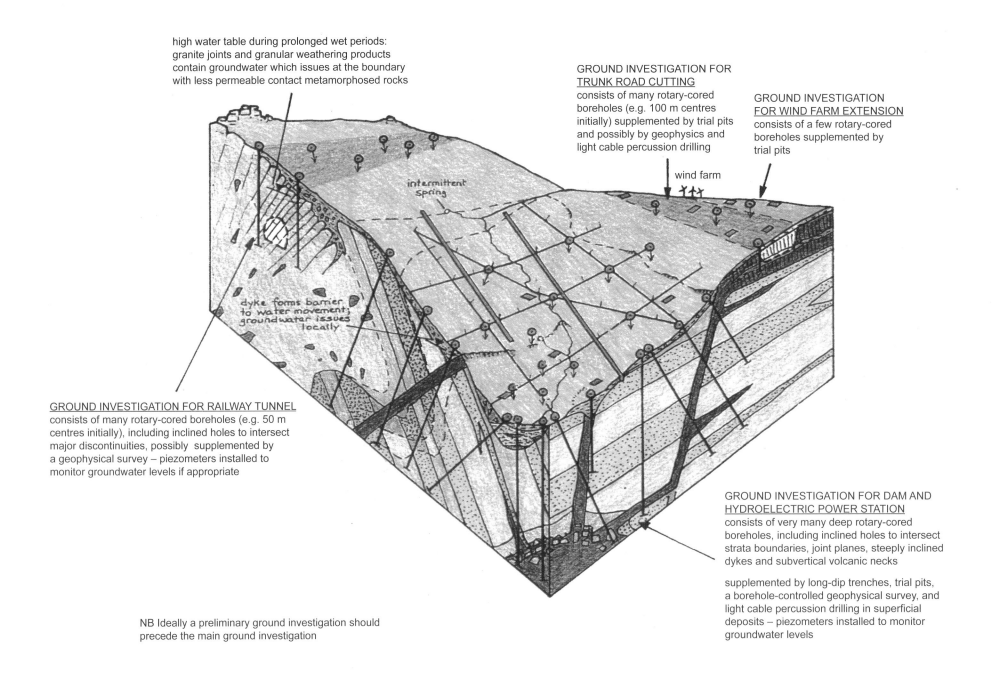

high water table during prolonged wet periods: granite joints and granular weathering products contain groundwater which issues at the boundary with less permeable contact metamorphosed rocks

GROUND INVESTIGATION FOR TRUNK ROAD CUTTING consists of many rotary-cored boreholes (e.g. 100 m centres initially) supplemented by trial pits and possibly by geophysics and light cable percussion drilling

GROUND INVESTIGATION FOR WIND FARM EXTENSION consists of a few rotary-cored boreholes supplemented by trial pits

wind farm

intermittent spring

dyke forms barrier to water movement; groundwater issues locally

GROUND INVESTIGATION FOR RAILWAY TUNNEL consists of many rotary-cored boreholes (e.g. 50 m centres initially), including inclined holes to intersect major discontinuities, possibly supplemented by a geophysical survey – piezometers installed to monitor groundwater levels if appropriate

GROUND INVESTIGATION FOR DAM AND HYDROELECTRIC POWER STATION consists of very many deep rotary-cored boreholes, including inclined holes to intersect strata boundaries, joint planes, steeply inclined dykes and subvertical volcanic necks

supplemented by long-dip trenches, trial pits, a borehole-controlled geophysical survey, and light cable percussion drilling in superficial deposits – piezometers installed to monitor groundwater levels

NB Ideally a preliminary ground investigation should precede the main ground investigation

*Figure 4.8 Idealized layout of ground investigations in steeply dipping tectonized strata.*

## *Idealized layout of ground investigations in steeply dipping strata (Figure 4.8)*

In the case of dipping beds, trenches and boreholes laid out parallel to the dip direction of the beds can reveal the complete stratigraphic sequence of a site. In addition, variations in the properties within particular dipping stratigraphic horizons with increasing depth below the ground surface can be obtained by sampling the particular horizon using a succession of down-dip boreholes. This can be achieved simply in unfaulted beds provided that the distance between the boreholes allows stratigraphic overlap of the recovered core to confirm the local stratigraphic sequence. Boreholes positioned up-dip penetrate lower in older stratigraphic levels than boreholes drilled to a similar depth but located down-dip – that is, information about lower stratigraphic levels can be obtained at reduced drilling costs by locating the drill hole up-dip rather than down-dip, providing that there is confidence in the extrapolation of this information.

Trenches laid out parallel to the strike of a bed can reveal lateral variations in the engineering properties along that bed. In addition, the vertical variation of the bed can be obtained by sampling from the top to the bottom of the bed at several locations. Where the dipping beds appear to be repeated on the basis of data obtained from two boreholes, a third borehole, or more, should be sunk to help interpret and correlate the sequence unless there is clear evidence of a fault causing the repetition of beds.

The correlation and interpretation of folded rocks may prove difficult with limited borehole information. Therefore observations should be made of the outcrop, noting any changes in dip or repetition/mirroring of the rock types. Again, supplementary boreholes sunk between two other boreholes where the same bed has different dips may aid in the interpretation of the structure. There are many variations in these examples depending on the complexity of the geology.

### SMALL-SCALE GEOLOGICAL FEATURES

The influence of small-scale geological features on geotechnical characteristics is fairly well documented. Rowe (1972) pioneered much of this work by relating the soil fabric to a range of sampling and testing procedures in engineering soils. Small-scale features in soils and sedimentary rocks are produced geologically by diagenesis (at and soon after the time of deposition, mainly during consolidation) and by tectonic events in any rock of any age, especially during plate collisions. They are therefore potentially significant in tectonized, steeply dipping strata.

For his work on sampling and testing, Rowe collected data from 35 sites, varying in age from the Ordovician to the Holocene, where the fabric of the clay soils was related to field and laboratory behaviour. These examples illustrated the inadequacy of conventional site investigations. Rowe advocated the description and recording of fabrics, and the use of these details and a knowledge of the overall geology and water levels in relation to the engineering problem, to decide the location, quality and size of specimens for fundamental laboratory tests. It was thus shown that it is not enough to simply use index tests alone for classification.

Advances in sampling and testing have developed in the last few decades, as exemplified by Vaughan (1994) in his Rankine Lecture, in which he stressed the links between the ground model behaviour and reality. Atkinson (2000), in his work on engineering soils, emphasized the value of carefully selected and well-designed laboratory work.

The significance of small-scale geological features, including texture and soil fabric, reaches beyond sampling and testing in the hands of a geologically skilled practitioner. Predictions can be made about the probable variability, characteristics and engineering properties of the material (soils and rocks), the characteristics of discontinuities, the origin of the material, the changes it has undergone and the probable suites of adjacent rocks and soils. Such features should be built into the initial geological and geotechnical site models. This will help in the interpretation and correlation of the subsur-face geometry of the geology, especially in complex situations, as the investigation proceeds. If the tectonic events have been severe – for example, producing metamorphic rock suites – many of the earlier small-scale features will have disappeared, only to be replaced by small-scale metamorphic features (see Figure 1.5).

As an example of the interpretation of small-scale features, Figure 1.8 shows a quarry in horizontally bedded sedimentary rocks. A trained and experienced geologist would know that the environment of deposition of these rocks before lithification would have been along the lines shown in Figure 1.7. This geologist could therefore predict the probable occurrence of the small-scale sedimentary details shown in insets A, B and C of Figure 1.8.

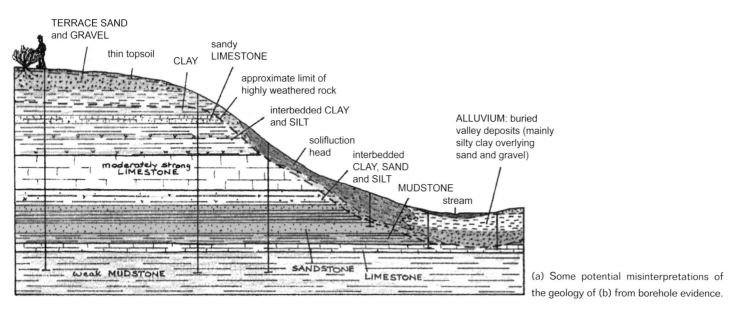

(a) Some potential misinterpretations of the geology of (b) from borehole evidence.

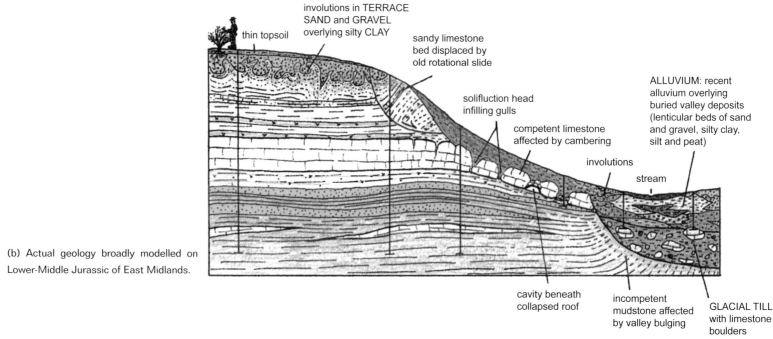

(b) Actual geology broadly modelled on Lower-Middle Jurassic of East Midlands.

*Figure 5.1 Problems in interpreting ground investigation information from the relict periglacial areas of southern Britain.*

# Part 5. Case histories and some basic ground characteristics and properties

## Problems in interpreting information from ground investigations in the periglacial areas of southern Britain (Figure 5.1)

Figure 5.1 is in two parts and shows some potential misinterpretations of the geology in the 'soft' rocks (usually weak to moderately strong sedimentary rocks, but can be outside these classifications) that occur in the southern parts of Britain. These are typically deposits of Jurassic and younger (Cretaceous and Tertiary) age. In Britain, these rocks have been subjected to periglacial activity at various times within the Quaternary. This activity has disturbed the near-surface strata (see Figures 2.1, 3.2, 3.5 and 5.9). In addition, these rocks may also have been disturbed in their upper parts by creep and other slope movements not necessarily related to periglacial activity and by warm temperate/subtropical forms of weathering (see Figures 2.2–2.5 and Bibliography).

In Britain, rocks of Jurassic and younger age have not been strongly disturbed by orogenic activity, although the distant effects of the Tertiary Alpine collisions have disturbed the rocks in the southernmost parts of Britain by folding and faulting. However, there has been neither intense folding nor metamorphism. Tertiary igneous activity was restricted mainly to northern Britain in association with the mid-Atlantic divergence (see Figures 1.2 and 1.3).

Figure 5.1a shows what might be interpreted at the main ground investigation stage from the careful logging of boreholes, combined with limited field mapping, at a project site in a relatively simple structural (i.e. beds not dipping) area of southern Britain. In this particular example, a similar interpretation might have been achieved by the following investigations: a study of local geological maps, memoirs and aerial photographs; a thorough walkover survey of the site and surrounding area; detailed engineering geological and geomorphological mapping; and the construction of large-scale geological cross-sections. It is assumed that borehole logging, fieldwork and data interpretation have all been carried out by experienced engineering geologists (Norbury, 2010).

Figure 5.1b shows the 'true' or 'as-found' geology, which includes relict periglacial involutions, rotational sliding, limestone strata displaced by cambering, and boulder glacial till beneath a varied sequence of buried alluvial deposits. It is likely that many of these features could have been revealed using a greater number of deeper boreholes, possibly including rotary-percussive probing, combined with trial pits, trenches and the use of geophysical techniques. Each unidentified feature may adversely affect construction operations and result in claims by the contractor for 'unforeseen' ground conditions.

Some important properties of the near-surface geology may only be revealed in excavations (e.g. pilot cuttings) during construction, whereas the deeper geology may never be completely known. Ideally, it will be the engineering geologist who interpreted the conditions at the ground investigation stage who has responsibility for updating the geological geomodel. If the observational method was built into the contract (Nicholson *et al.*, 1999), any required design modification can be implemented as construction proceeds.

Examples of periglacial activity and other forms of near-surface disturbance include the following.

- Frost action, including cryoturbation and other forms of slight to severe churning of the ground, particularly ice wedges and pingos. These are all generally related to the active layer (see Figure 5.9). Chalk is especially susceptible to this kind of activity. Table 5.1.1 (after Bell and Culshaw, 2005) describes the chalk weathering grades, which differ significantly from 'normal' weathering grades (see Figure 2.3).

- Solifluction flows and other forms of mass movement of slope debris previously related to the active layer.

- The formation of gulls (very large cracks that open more widely near the ground surface) in cambered (downwards bending or draping) ground primarily related to periglacial conditions.

- The deposition of alluvial and other stratified superficial soils, including metastable wind-blown loess ('brick earth' in Britain) and cover sands, both of which can be metastable and locally extensive.

There were long periods of hot and wet climates in Britain in the geological past, particularly during the Tertiary, leading to a residual soil by weathering of the upper few metres of the ground. The results of such weathering may still be preserved and can be locally relatively common, especially in south-west Britain. During the Tertiary period (and before), 'Britain' was south of where it is now and slowly moved northwards through a hotter and wetter climatic belt to its present cooler location.

Table 5.1.1 Description and grading of the Middle and Upper Chalk for engineering purposes.

| Grade | Structure | Colloquial description of grade | Definitions of grade | | Typical features of grade | | Word order for description |
|---|---|---|---|---|---|---|---|
| | | | *Comminuted chalk matrix (%)* | *Coarser fragments (%)* | *Weathering of coarser fragments* | *Strength of coarser fragments* | |
| cVI | Structureless chalk, bedding and jointing absent (cVI, cV) | Putty chalk with small lumps | >35 | <65 | Moderately, highly or completely weathered | Very weak or weak | Structureless chalk: soil strength of matrix; colour of matrix; nature of matrix material; amount of fragments; presence and nature of flints; other features (Grade VI) |
| cV | | Chalk lumps in comminuted matrix | <35 | >65 | Moderately or highly weathered | Very weak or weak | Structureless chalk: angularity and size of fragments; colour of fragments; weathering of fragments; strength of fragments; amount of matrix; nature of matrix material; presence and nature of flints; other features (Grade V) |
| | | | *Fracture spacing (mm)* | *Fracture width (mm)* | *Material weathering* | *Material strength* | *Colour: rock material, weathering chalk* |
| cIV | In situ structured chalk, with bedding and jointing (cIV, cIII, cII, cI) | Rubbly chalk | Extremely closely to very closely spaced <60 | Open or infilled >5 | Moderately or highly weathered | Very weak or weak | Rock material; strength; discontinuity type; discontinuity spacing; discontinuity width and nature of infill if appropriate; discontinuity orientations (*in situ* observations only); presence and nature of flints; other features (give Grade) |
| cIII | | Rubbly to blocky chalk | Closely spaced 60–200 | Open or infilled <3 | Slightly or moderately weathered | Weak or moderately weak | |
| cII | | Blocky chalk | At least medium spaced >200 | Tight and clean | Fresh or slightly weathered | Weak or moderately weak | |
| cI | | Brittle and massive chalk | At least medium spaced >200 | Tight and clean | Fresh or slightly weathered | Moderately weak or moderately strong | |

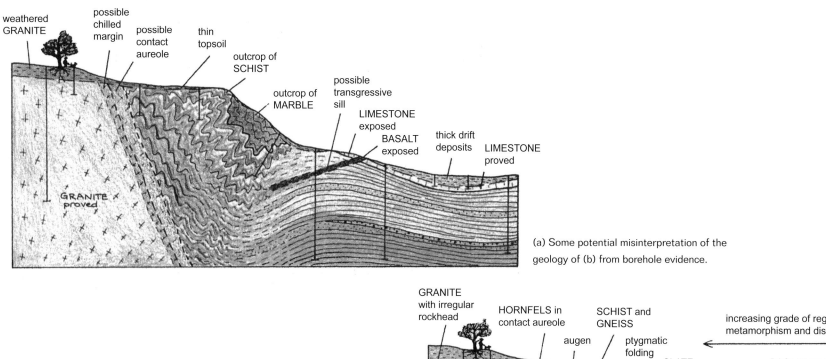

weathered GRANITE

possible chilled margin

possible contact aureole

thin topsoil

outcrop of SCHIST

outcrop of MARBLE

possible transgressive sill

LIMESTONE exposed

BASALT exposed

thick drift deposits

LIMESTONE proved

GRANITE proved

(a) Some potential misinterpretation of the geology of (b) from borehole evidence.

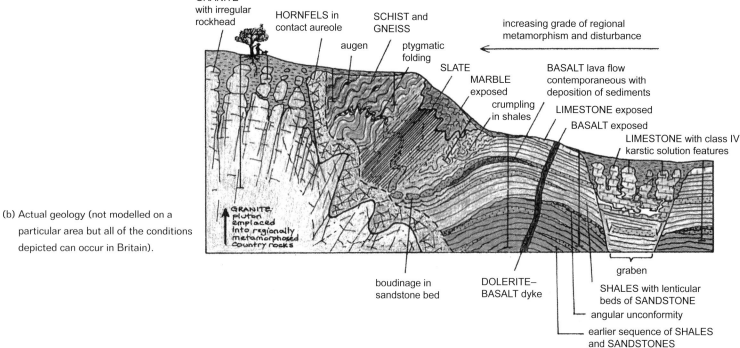

GRANITE with irregular rockhead

HORNFELS in contact aureole

SCHIST and GNEISS

augen

ptygmatic folding

SLATE

MARBLE exposed

crumpling in shales

increasing grade of regional metamorphism and disturbance

BASALT lava flow contemporaneous with deposition of sediments

LIMESTONE exposed

BASALT exposed

LIMESTONE with class IV karstic solution features

GRANITE pluton emplaced into regionally metamorphosed country rocks

(b) Actual geology (not modelled on a particular area but all of the conditions depicted can occur in Britain).

boudinage in sandstone bed

DOLERITE–BASALT dyke

SHALES with lenticular beds of SANDSTONE

angular unconformity

earlier sequence of SHALES and SANDSTONES

graben

*Figure 5.2 Problems in interpreting ground investigation information in structurally complex regions.*

## Problems in interpreting information from ground investigations in structurally complex regions (Figure 5.2)

Figure 5.2 is also in two parts and is of a region that is relatively complex structurally. Figure 5.2a shows some potential misinterpretations that might occur during a ground investigation and Figure 5.2b shows the actual geology found during excavations.

When fresh, rocks in these complex regions are typically moderately to extremely strong (see Table 5.3.1), although exceptions do exist. Such rock is loosely called 'hard' (meaning strong). Logging needs to be carried out by experienced engineering geologists (Norbury, 2010). Structurally complex ground typically occurs in the northern and western areas of Britain and in many other locations around the world. Such rocks in Britain are of Permo-Triassic age or older and have been through at least one important orogenic event. Older rocks are more likely to have been affected by several orogenic episodes and hence be structurally very complex. Rocks in Britain can be as old as the Precambrian. During an orogenic episode, the original rocks may have been heated and subjected to pressure, producing metamorphic rocks, and may also have been subject to strong folding, faulting and shearing without necessarily being metamorphosed. They may also have been invaded, at any time, by igneous magma and therefore may include volcanic sequences, small intrusions and possibly plutonic emplacements. Igneous activity last occurred in Britain during the Tertiary (mainly northern Britain) (see Figures 1.4, 1.5, 1.10 and 2.2).

Figure 5.2a shows what might be interpreted from the careful logging of boreholes combined with limited field mapping at the project site by experienced engineering geologists at the main ground investigation stage. This interpretation recognized important features such as the granite intrusion (perhaps part of a batholith) on the left-hand side, with a possible contact aureole of country rocks baked by heat during emplacement of the granite. Regionally metamorphosed gneisses, schists and marble beyond the aureole

were distinguished from laterally equivalent unmetamorphosed shales, sandstones and an upper thin limestone bed. A basic igneous intrusion of unknown form has been revealed and the presence of thick Quaternary deposits (alluvium and colluvium) recognized on the valley floor on the right-hand side of the figure. It is possible that a broadly similar interpretation of the upper weathered strata could have been achieved from a desk study, walkover and field mapping, but the changes in rock type and structure with depth could not have been predicted.

Figure 5.2b shows the 'true' or 'as found' geology, which is significantly more complicated than the model produced at the main ground investigation stage. The granite, which has been subject to warm climate weathering during the Tertiary, has a very irregular rockhead with large corestones. The aureole is composed of hornfels, typically a very strong, brittle rock, and the regionally metamorphosed strata beyond it include slates at depth and minor structural features such as augen, ptygmatic folding and boudinage. The sedimentary sequence contains lenticular sandstones and a lava flow. There is also a subvertical basic igneous dyke. Very importantly, a graben structure on the valley floor consists of preserved limestone that has been subjected to significant karstic solution.

Although a more intensive ground investigation would have disclosed more of the underlying geology, some important near-surface characteristics would probably only have been revealed in excavations during construction. The importance of using the observational method to refine the geological model and to make design modifications as construction proceeds is again emphasized.

The features shown in Figure 5.2 are examples of some of the basic situations that may arise, but there are many more that also require interpretation by an experienced geologist. In addition to the changes shown, such 'hard' rock areas usually display increased rock fracturing towards the ground surface as a result of glacial and periglacial action, or perhaps daily hot/cold temperature changes. The removal of overburden by erosion creates additional rock fractures when the rock relaxes after confinement. These are known as 'stress relief'

or 'unloading' fractures and are typically subparallel to the ground surface (whether flat ground or a mountainside). In the short term, similar fractures can develop inside a rock face created by excavation or quarrying. Although drilling aims for 100% core recovery, faults may have sheared or gouged the rock to produce weak zones that are difficult to recover as core. Deep weathering under warmer and wetter conditions may also have produced weaker, more weathered rocks overlain by stronger rock. During the course of its geological history, the top of the bedrock below the overburden may well have become convoluted (e.g. by tectonism). The late-stage geomodel must aim to include all of these features.

An escarpment formed by a gently dipping bed of strong limestone has a steep scarp face rising above extensive, active scree slopes.

A steep mountain face is formed in structurally complex metamorphic gneiss with dipping banding and foliation, but the rock is strong enough to require no support within the road tunnel, mainly because it has a very low density of fractures.

A horizontal sequence of sedimentary rocks forms the upper half of a cliff, above an unconformity on top of structurally complex basement rocks.

A massive landslide came away from the steep mountain face that was formed of steeply folded limestone overlying a major thrust plane; failure was aided by glacial over-steepening of the face, mining of coal beneath its toe and input of water when ground ice thawed in the spring.

Multiple landslides have left a complex bank of rotated slices of rock that underlie the chaotic topography below the cliff formed by the back scar.

Multiple phases of tight folding in metamorphic rocks can be on all scales, so this view could be centimetres or hundreds of metres across.

Boudinage formed in the upper, older, dolerite dyke as it was stretched and "necked" by plastic deformation at high pressure and temperature, so the intrusion now varies from thick to negligibly thin.

A banded sequence of easily eroded clay has zig-zag folds clearly exposed at outcrop, but very difficult to identify just from a set of widely spaced boreholes.

[above] Complex folding of thinly bedded shales and limestones includes sections where beds are nearly vertical, all of which would be very difficult to interpret from just a few boreholes and would require field mapping for a full assessment.

[left] Within a mountain fold belt, a cliff face more than 200 metres high exposes major faults through complex folds with vertical and horizontal beds.

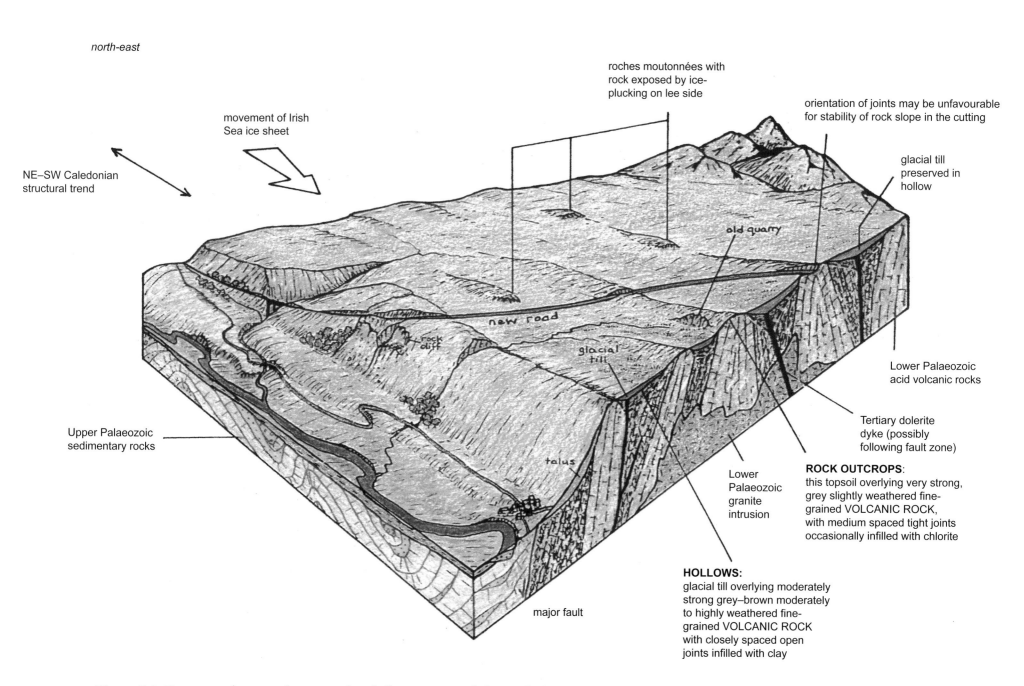

north-east

movement of Irish
Sea ice sheet

roches moutonnées with
rock exposed by ice-
plucking on lee side

orientation of joints may be unfavourable
for stability of rock slope in the cutting

glacial till
preserved in
hollow

NE–SW Caledonian
structural trend

old quarry

new road

rock
cliff

glacial
till

Lower Palaeozoic
acid volcanic rocks

Upper Palaeozoic
sedimentary rocks

talus

Tertiary dolerite
dyke (possibly
following fault zone)

Lower
Palaeozoic
granite
intrusion

**ROCK OUTCROPS**:
this topsoil overlying very strong,
grey slightly weathered fine-
grained VOLCANIC ROCK,
with medium spaced tight joints
occasionally infilled with chlorite

major fault

**HOLLOWS**:
glacial till overlying moderately
strong grey–brown moderately
to highly weathered fine-
grained VOLCANIC ROCK
with closely spaced open
joints infilled with clay

*Figure 5.3 Conceptual route of new road in hilly terrain and the underlying relations between landforms and geology: walkover 1, rocks.*

## Route of a new road in hilly terrain: walkover 1, rocks (Figure 5.3)

Figure 5.3 and its companion Figure 5.4 should be worked together. They are both idealized views loosely based on a real case study in Britain, although there have been some alterations in the preparation of the pictorial block diagrams. Figures 5.3 and 5.4 are the results of separate walkover surveys by two different engineering geologists, each working for a construction contractor bidding for a road contract in hilly, generally rocky, terrain. The geologists each prepared a small desk study and then walked or drove over the proposed route; the ground investigation data issued to all the bidding contractors was made available to both geologists. The figures illustrate the interpretations that each geologist made of the subsurface characteristics of the ground, which were important in helping the contractors to decide the method of excavation for the cuttings. The geologist of Figure 5.4 overestimated the amount of very strong rock in the cuttings; the geologist of Figure 5.3 estimated a lesser amount of strong rock because significant weathering was spotted. The contractor of Figure 5.3 chose to rip and hammer, whereas the contractor of Figure 5.4 chose to blast – a bigger bid price was therefore put in and the contract was lost.

### WALKOVER SURVEY: ROCK

The figures in Part 4 illustrate studies for large ground investigations and Figure 4.3, in particular, discusses the walkover survey stage. Throughout the ground investigation, soil and rock descriptions are made during the walkover from natural exposures, from cuttings and excavations, and from disturbed and undisturbed samples taken from pits and boreholes. Trained and experienced engineering geologists (and geotechnical engineers) make continual judgements as they walk or drive around a proposed site. Good practitioners may well get right, or nearly right, most of their judgements of the engineering characteristics of soils and their distribution from these observations, including the characteristics and configurations of the near-surface rocks. Groundwater conditions and the subsurface geological structures are more difficult to predict from surface observations. All of these judgements require verification during the intrusive and laboratory stages of the ground investigation and, as previously noted, the geology may only be revealed fully when construction begins.

There are many schemes for the description of soils and rocks (Norbury, 2010). These descriptions are often based on the particle size and plasticity of soils or the crystal structure and other mass characteristics of rocks. They generally contain information on some or all of the following characteristics, as appropriate:

- the field strength or compactness;
- the structure (e.g. bedding, folding, shearing, discontinuities) and state of weathering;
- the colour;
- the particle or crystal shape and composition;
- the SOIL NAME, based on particle size, and the ROCK NAME, based on the rock type; and
- reference to inclusions and any other observations.

Tables 5.3.1–5.3.3 give common examples of the sort of field observations that engineering geologists make in rocky terrains. Reference should also be made to more comprehensive textbooks and codes of practice (see Bibliography, Group B books; Ulusay and Hudson, 2007; Norbury, 2010). These tables will be subject to frequent updating/changes when codes are revised; when reporting fieldwork, the particular reference to the classification used must be given.

Table 5.3.4 gives some typical properties of common rocks for which an experienced engineering geologist or geotechnical engineer would have a good working understanding. Again, there are many more properties of significance to design and performance that could be evaluated, if required, in the laboratory and by field tests. Such tests would also verify the geologist's field estimates; see Head (2006) and Head and Epps (2011, 2014) for detailed descriptions of laboratory testing. It is worth emphasizing that the rock characteristics and probable foundation performance vary significantly within one rock type, depending on the mineralogy and the state of

Table 5.3.1 Classification of rock strength.

| Strength | Field strength (MPa) | Field properties |
|---|---|---|
| Very weak | 1–5 | Crushes between fingers |
| Weak | 5–25 | Breaks easily by hand |
| Medium strong | 25–50 | Breaks with a single hammer blow |
| Strong | 50–100 | Requires more than one hammer blow to break |
| Very strong | 100–250 | Chipped by heavy hammering |
| Extremely strong | >250 | Rings when hammered |

Table 5.3.2 Shapes of joints and bedding-bound natural blocks.

| Shape | Dimensions |
|---|---|
| Blocky | Equidimensional |
| Tabular | Thickness much less than length or width |
| Columnar | Columnar and elongated; largest dimension greater than twice any of the others |
| Irregular | No axes of similar length |
| Flaggy | Smallest dimension 20–60 mm; other dimensions at least twice that of smallest |
| Slaty or shaly | Smallest dimension <20 mm; other dimensions at least twice that of smallest |

Table 5.3.3 Basic spacing of discontinuities.

| Description | Spacing |
|---|---|
| Very widely spaced | >2 m |
| Widely spaced | 2 m to 600 mm |
| Medium spaced | 600–200 mm |
| Closely spaced | 200 60 mm |
| Very closely spaced | 60–20 mm |
| Extremely closely spaced | <20 mm |

weathering and fracturing (e.g. joints and faults). These are taken into account by competent experienced practitioners (see Figures 1.10 and 2.3).

Table 5.3.4. Typical mechanical properties of some common rocks.

| Rock type | Dry density (t/m3) | Porosity (%) | Dry UCS (MPa) | | Saturated UCS (MPa) [b] | Modulus of elasticity (GPa) | Tensile strength (MPa) | Shear strength (MPa) | Friction angle (Φ°) |
|---|---|---|---|---|---|---|---|---|---|
| | | | *Range* [a] | *Mean* | | | | | |
| *Igneous/plutonic rocks* | | | | | | | | | |
| Basalt | 2.9 | 2 | 100–350 | 250 | | 90 | 15 | 40 | 50 |
| Granite | 2.7 | 1 | 50–350 | 200 | | 75 | 15 | 35 | 55 |
| *Metamorphic rocks* | | | | | | | | | |
| Hornfels | 2.7 | 1 | 200–350 | 250 | | 80 | | | 40 |
| Marble | 2.6 | 1 | 60–200 | 100 | | 60 | 10 | 32 | 35 |
| Gneiss | 2.7 | 1 | 50–200 | 150 | | 45 | 10 | 30 | 30 |
| Schist | 2.7 | 3 | 20–100 [c] | 60 | | 20 | 2 | | 25 |
| Slate | 2.7 | 1 | 20–250 [c] | 90 | | 30 | 10 | | 25 |
| *Clastic rocks* | | | | | | | | | |
| Greywacke | 2.6 | 3 | 100–200 | 180 | 160 | 60 | 15 | 30 | 45 |
| Carboniferous sandstone | 2.2 | 12 | 30–100 | 70 | 50 | 30 | 5 | 15 | 45 |
| Triassic sandstone | 1.9 | 25 | 5–40 | 20 | 10 | 4 | 1 | 4 | 40 |
| *Carbonate rocks* | | | | | | | | | |
| Carboniferous limestone | 26 | 3 | 50–150 | 100 | 90 | 60 | 10 | 30 | 35 |
| Jurassic limestone | 2.3 | 15 | 15–70 | 25 | 15 | 15 | 2 | 5 | 35 |
| Chalk | 1.8 | 30 | 5–30 | 15 | 5 | 6 | 0.3 | 3 | 25 |
| *Mudrocks* | | | | | | | | | |
| Carboniferous mudstone | 2.3 | 10 | 10–50 | 40 | 20 | 10 | 1 | | 30 |
| Carboniferous shale | 2.3 | 15 | 5–30 | 20 | 5 | 2 | 0.5 | | 25 |
| Cretaceous clay | 1.8 | 30 | 1–4 | 2 | | 0.2 | 0.2 | 0.7 | 20 |
| *Organic* | | | | | | | | | |
| Bituminous coal | 1.4 | 3 | 3–30 | 20 | | 5 | 2 | | |
| *Evaporites* | | | | | | | | | |
| Gypsum (sulphate) | 2.2 | 5 | 20–30 | 25 | | 20 | 1 | | 30 |
| Salt (chloride) | 2.1 | 5 | 5–20 | 12 | | 5 | | | |

[a] Probably contains results from weathered and unweathered rocks and rocks with incipient cracks, i.e. a large spread of results.

[b] Samples from dried-out core can give significantly different results from rock with in situ moisture.

[c] Will contain results from samples tested parallel, or normal, to the natural planes of weakness, i.e. a large spread of results.

Rebuiding the North Wales Coast Road was a major engineering project that had some innovative and successful features of engineering geology and geotechnics. These images do not relate to the text or features shown in the block diagrams on the adjacent pages, but show engineering works carried out on broadly similar terrain.

Early excavation work shows the very constricted site, with the live railway below.

Parts of the coastal strip were an undercliff formed by the toes of landslides, and one area was stabilised by placing the road immediately behind a massive sea wall that was sufficiently massive to form an anchor weight over the rising part of the slip surface.

At Towyn, sea defences have 4-tonne blocks of limestone placed individually by a grab.

The tunnel under the River Conway is a submerged tube that has approaches in deep cuttings with retaining walls; a thickened concrete base was made with a dense aggregate to hold the structure down and prevent it floating upwards where it lies beneath the water table.

[above] The Pen y Clip Tunnel required massive steel arches for roof support through zones of fractured and deeply weathered rock.

[left] The Penmaenbach Tunnel has rock anchors in the exposed face above the portal to prevent fractures opening by stress relief.

[right] Parts of the Rhuallt cutting had individual designs of anchors, rock bolts and dental masonry in dipping and faulted greywackes.

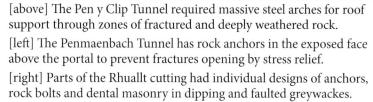

[left] Approaches to the Pen y Clip Tunnel have steep cut faces, where the unstable nature of the strong but densely fractured rock required continuous support by top-down construction of rows of anchored concrete panels.

[right] Split carriageways that contour round steep hillsides above the shoreline are each supported by vertical retaining walls; these are held in place by rock anchors, some of which had to be 40 metres long to reach through the weathered rock into stable ground.

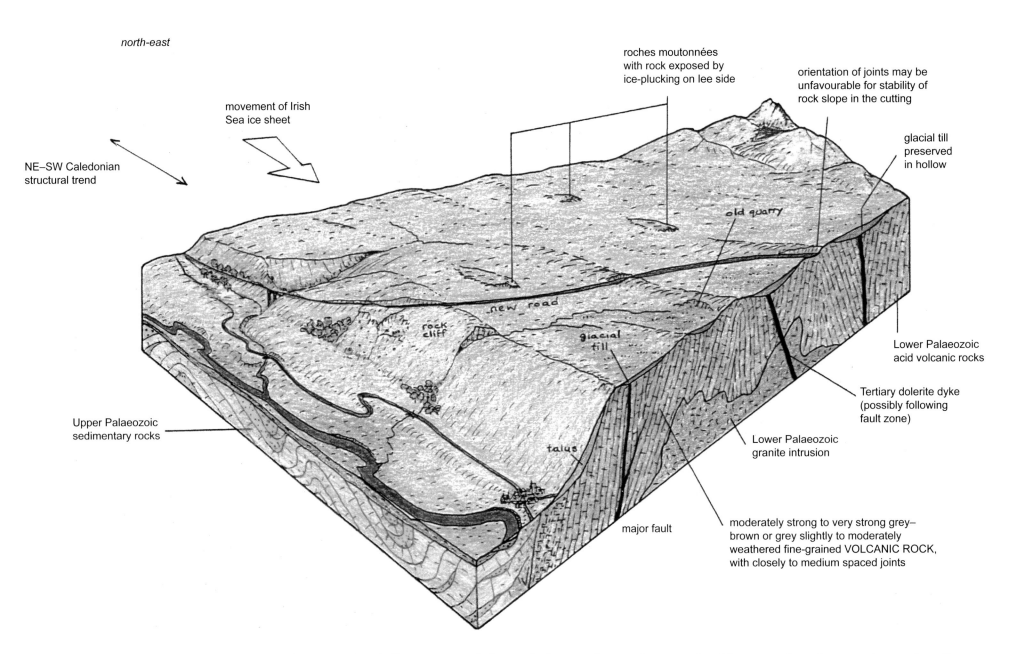

north-east

movement of Irish
Sea ice sheet

NE–SW Caledonian
structural trend

roches moutonnées
with rock exposed by
ice-plucking on lee side

orientation of joints may be
unfavourable for stability of
rock slope in the cutting

glacial till
preserved
in hollow

old quarry

new road

rock
cliff

glacial
till

Lower Palaeozoic
acid volcanic rocks

Upper Palaeozoic
sedimentary rocks

talus

Tertiary dolerite dyke
(possibly following
fault zone)

Lower Palaeozoic
granite intrusion

major fault

moderately strong to very strong grey–
brown or grey slightly to moderately
weathered fine-grained VOLCANIC ROCK,
with closely to medium spaced joints

*Figure 5.4 Conceptual route of new road in hilly terrain and the underlying relations between landforms and geology: walkover 2, soils.*

## Route of new road in hilly terrain: walkover 2, soils (Figure 5.4)

Figure 5.4 should be worked with Figure 5.3. Both figures are of walkover surveys of the same site by two engineering geologists, each working for a separate contractor. The text of Figure 5.3 for convenience discusses the strong rocks, whereas the text of Figure 5.4 discusses the superficial materials and engineering soils.

### WALKOVER SURVEY: SOILS

Tables 5.4.1–5.4.3 are examples of the basic field observations that engineering geologists and geotechnical engineers might make on superficial materials and engineering soils and, if appropriate, weak sedimentary rocks. There are considerably more selected characteristics than given here and reference should be made to more comprehensive textbooks and Codes of practice (see Bibliography, Group B books; Norbury, 2010). These tables will need to be updated/changed when codes are revised.

The figures in Part 4 develop a staged sequence of studies for a large ground investigation. Figure 4.3 discusses the walkover survey stage. The figures in Part 2 discuss near-surface weathering and other changes that have a strong influence on the engineering performance of the soils (and rocks) encountered. Figures 4.7 and 4.8 give further examples of common ground in the UK, including superficial and engineering soils.

Table 5.4.1 is a much simplified presentation of the principal Unified Soil Classification Soil Groups and the range of sizes for each group (BS 5930:1999; Norbury, 2010, especially Chapters 2–8). The basic angle of internal friction (Φ) and the Atterberg properties are also given to help understand the probable field and design behaviour. It is emphasized that the values quoted are only 'typical' and that significant variations could occur and would be found by good laboratory testing. The angle of internal friction Φ is a result of the structural roughness between grains and is higher for cohesionless clastic soils (e.g. silts, sands, gravels) composed of fragments

Table 5.4.1 Unified Soil Classification: some basic characteristics.

| Group | Symbol | Grain size (mm) | Typical values | | |
|---|---|---|---|---|---|
| | | | LL | PI | Φ |
| Gravel | G | 2–60 | – | – | >32 |
| Sand | S | 0.06–2 | 40 | – | >32 |
| Silt (low plasticity) | ML | <0.002–0.06 | 60 | 15 | 32 |
| Clayey silt | MH | <0.002–0.06 | 35 | 30 | 25 |
| Clay (low plasticity) | CL | <0.002 | 70 | 20 | 28 |
| Plastic clay | CH | <0.002 | | 45 | 19 |
| Organic | O | – | | | <10 |

of pre-existing rocks than for cohesive soils (e.g. clays). In cohesive soils, Φ is synonymous with the angle of shearing resistance.

Natural clays are typically plastic. With varying water contents, a clayey soil may be solid, plastic or liquid, as measured by its *Atterberg limits*. These are index tests carried out on remoulded samples in which the natural fabric of the soil has been destroyed. The *plastic limit* (PL) is the minimum moisture content needed to roll the soil into a cylinder of 3 mm diameter. Soil at its PL has a shear strength of about 100 kPa. The *liquid limit* (LL) is the minimum moisture content at which the soil flows under its own weight. The important *plasticity index* (PI) (where PI = LL – PL) is the change in water content required to increase the strength 100 times and is the range of water contents at which the soil behaviour is plastic or sticky. Soils with a high PI are generally less stable and have a large swelling potential.

Table 5.4.2 gives some basic cohesive soil properties for design and construction, allied to simple, but very important, field observations. All soils fail in shear; the *shear strength* is a combination of the cohesion and the internal friction expressed by the Coulomb failure envelope. This is a subject for discussion in more specialized textbooks (see Bibliography, Group B books). *Cohesion*, derived from molecular attraction forming inter-particle bonds, is significant in clays, but zero in clean sands. The shear strength = (cohesion + normal stress) ×

tan F. *Normal stress* (stress is the force that produces deformation in a body) is critical to the shear strength, but the *pore water pressure* (pore water is the water in the space between particles) carries part of the overburden load on the soil and therefore subtracting the pore water pressure from the normal stress leaves the effective stress. The concept of effective stress is of vital importance in soil mechanics, in particular where structures are stressed over long periods (e.g. ground slope instability) (see Bibliography, Group B books; Head, 2006; Head and Epps, 2011, 2014). The drainage progress of a loaded clay (i.e. one forming the foundation to a structure) is critical because any increase in pore water pressure may lead to failure. This is particularly significant in new excavations and embankments that change the drainage arrangements.

Table 5.4.3 gives the common field identification and strength characteristics of loose to densely packed sands. These properties are variable and will need to be confirmed in the laboratory. Sand and gravel soils have no cohesion except that derived from any clay matrix or mineral cement between the grains and from water suction. Sands can stand as steep slopes when moist as a result of the negative pore pressure (critical in building sandcastles), but will not stand when dry or saturated. The strength, slope stability and bearing capacity all derive from internal friction and Φ for granular soils ranges from about 30 to 45°. This angle increases with coarser grades, the packing density and grain angularity. Settlement in sands is typically small and rapid and is not usually considered, except on very loose sands and artificial fills. Properties are commonly estimated in the field from the *standard penetration test* (SPT N values) made with a percussion boring rig during the preliminary or main investigations. When using a 64 kg hammer dropped 760 mm to drive a standard tube 300 mm, N is given by the number of blows required. Frequently derived basic properties from the corrected blow count N are the relative density and the value of Φ.

Table 5.4.2 Some typical properties of cohesive fine soils.

| Examples | State | Consistency of clays | Liquidity | SPT, N | CPT (MPa) | Cohesion (kPa) | Mv (m²/ MN) | ABP (kN/ m²) |
|---|---|---|---|---|---|---|---|---|
| Alluvial clays, Tills Cainozoic and Mesozoic Engineering Clays (e.g. London Clay, Oxford Clay) | Very soft | Fingers easily pushed in up to 25 mm | 0 | <2 | – | 0–20 | – | – |
| | Soft | Finger pushed in up to 10 mm | >0.5 | 2–4 | 0.3–0.5 | 20–40 | >1.0 | <75 |
| | Firm | Thumb makes impression easily | 0.2–0.5 | 4–8 | 0.5–1 | 40–75 | 0.3–1.0 | 75–150 |
| | Stiff | Indented slightly by thumb | −0.1 to 0.2 | 8–15 | 1–2 | 75–150 | 0.1–0.3 | 150–300 |
| | Very stiff | Indented by thumbnail | −0.4 to −0.1 | 15–30 | 2–4 | 150–300 | 0.05–0.1 | 300–600 |
| | Hard | Scratched by thumb nail | Less than −0.4 | >30 | >4 | >300 | <0.005 | >600 |

Cohesion $c$ is equivalent to short-term shear strength.

APB = allowable bearing pressure; CPT = cone penetration test; $Mv$ = coefficient of compressibility.

$\Phi$ varies with density in cohesionless soils (Table 5.4.3).

Table 5.4.3 Typical properties of sands.

| Packing | Identification | RD | SPT | CPT | Φ | SBP |
|---|---|---|---|---|---|---|
| Very loose | – | <0.2 | <4 | <2 | <30 | <30 |
| Loose | Can be dug by spade, 50 mm peg easily driven in | 0.2–0.4 | 4–10 | 2–4 | 30–32 | 30–80 |
| Medium dense | – | 0.4–0.6 | 11–30 | 4–12 | 32–36 | 80–300 |
| Dense | Needs pick for excavation, 50 mm peg hard to drive in | 0.6–0.8 | 31–50 | 12–20 | 36–40 | 300–500 |
| Very dense | – | >0.8 | ≥50 | ≥20 | ≥40 | ≥500 |
| Slightly dense | Visual examination; pick removes soil in lumps that can be abraded | >0.8 | >50 | Depends on cementing | | |

RD = relative density; SPT = $N$ values; CPT = end resistance in MPa; Φ = angle of internal friction; SBP (safe bearing) in kPa for foundations 3 m wide with settlements <25 mm.

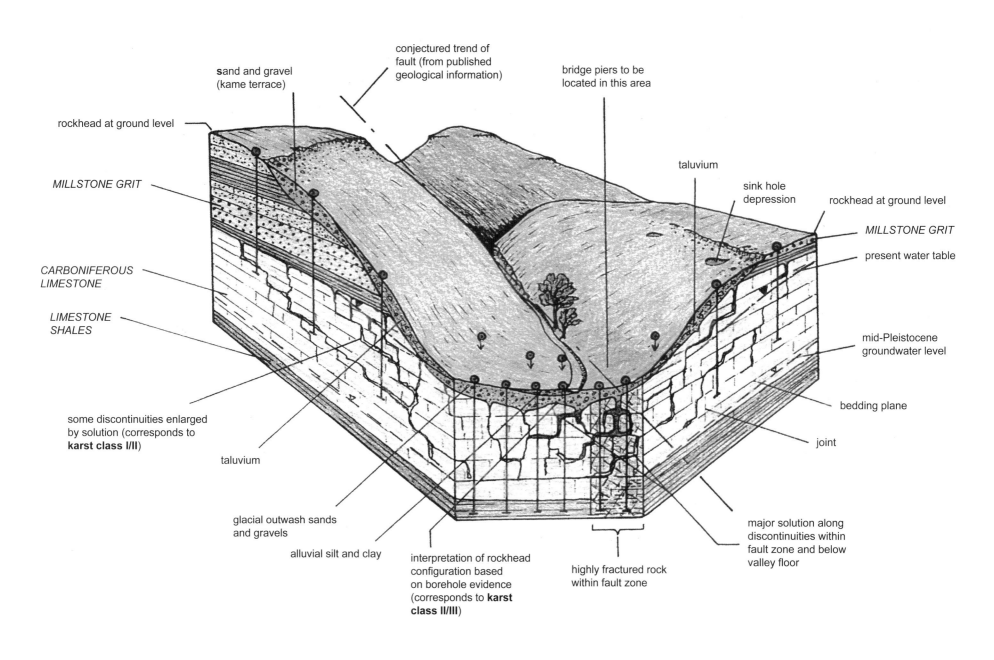

*Figure 5.5 Foundation conditions in limestone bedrock interpreted from the ground investigation for a river crossing.*

158

## Karst conditions in limestone bedrock interpreted from the ground investigations for a river crossing (Figure 5.5)

Figure 5.5 should be worked with Figure 5.6. Both figures are loosely based on a real case history in Britain of a large bridge crossing in limestone terrain with karstic features. An opportunity has been taken in the figures to review briefly the subsurface ground alterations created by limestone solution (karst).

Karst develops where the solution of rocks predominates over any other weathering and erosion processes. The key surface features of karst terrain include dry valleys and closed depressions (sink holes or dolines) of varying sizes, pinnacled rockhead, bare limestone pavement in colder regions, and cones and towers in tropical karst, together with small-scale solution sculpting of exposed rock anywhere. Near-surface karst types include bare or soil-covered rock where the rockhead is less than 1 or 2 m below the surface. Mantled karst with solution-eroded topography is covered by 10 m or more of unconsolidated deposits. These deposits may include the insoluble residues of dissolved limestone or the insoluble remains of strata that once overlaid the limestone (Waltham and Fookes, 2003).

Underground drainage is channelled into caves, conduits (pipes) or fissures formed by solution. Diffuse flow through fractures and smaller voids can occur in chalks, but evolves into conduit flow in the stronger limestones. Such terrains are of considerable significance to load-bearing engineering structures and in the understanding of local drainage systems for engineering projects.

Carbonates (principally calcium carbonate, such as limestones) are easily the most common of the soluble rocks that form karst. The largest caves and the most rugged surface karst landforms are formed in older, stronger limestones and also in the less common, but stronger, dolomitic rocks (magnesium carbonates) and take hundreds of thousands of years to form. Other soluble rocks that form karst features include gypsum (calcium sulphate), which typically has thinner beds, a wider development of interstitial karst and a great number of connecting pipes. However, these rocks do not develop to the cone or tower karst stage (see Figure 5.6). Gypsum and other sulphates are dissolved more quickly than the carbonate family and are commonly associated with arid or semi-arid regions. The solution of gypsum is usually relatively fast and can even occur over engineering time.

### ENGINEERING HAZARDS

Engineering hazards on karst terrain can include the following.

- *Problem foundation conditions.* These may occur in areas with highly irregular, pinnacled rockhead relief and adjacent boreholes may reach rock at depths varying by 2 m or more. It is important to explore the area thoroughly before constructing end-bearing piles.

- *Ground voids.* These are caves and fissures that are empty, water-filled or filled with 'soft' sediments. They can cause rare, isolated ground collapses under structural loads if not previously identified by diligent investigation in critical locations.

- *Sink hole collapses.* These are formed by the down-washing of the soil cover infilling an existing sink hole into fissures and caves. These are the most widespread geohazards in soil-covered karst. All new sink hole collapses are the result of influxes of water, so they may be caused by rainstorms or fractured pipelines, although it is considered that most are caused by modified or uncontrolled surface drainage on construction projects and are therefore largely avoidable. In Britain, this all too often results in houses on new developments rapidly falling into new sink hole collapses.

- *Reservoir leakage.* This is only remedied by grouting on a massive scale. In China, for example, one-third of the 5000 reservoirs impounded on limestone karst suffer from significant continual leakage. The rapid solution of gypsum means that reservoirs are generally not practicable on gypsum karst.

The investigation of karstic terrain always causes problems. Geophysical techniques are commonly not successful and therefore should be carried out under the guidance of a consultant geophysicist. Drilling techniques often have difficulty in core recovery as a result of the sudden occurrence of unpredictable voids of varying sizes, commonly (but not always) filled with loose or poorly consolidated carbonate sediments (see the ground investigation column in Table 5.6.1; each investigation in this column is directly related to the class of karst).

A commentary is given in the text of Figures 4.2–4.6 and outlines the five key stages of a ground investigation, as an example of how the karst class was estimated at each stage. This indicates the level of investigation needed as the ground investigation proceeds through each of its five stages.

An abandoned quarry in strong limestone, with sawn faces that revealed the extent of solution fissures, both open and filled with brown, terra rosa clay, in a karst of class kIII.

Limestone pinnacles exposed on a building site in class kIV karst, before digging out the clay soil, breaking off the pinnacle tops and filling the fissures with broken rock.

[below right] A large subsidence sinkhole in glacial till 10 metres thick, with limestone exposed where a small stream now drains into the site. Formed in a long series of small events, a slice of soil on the far side had recently slumped into the open fissure below.

[below left] A road cutting five metres high exposing buried sinkholes filled with clay in karst of class kII.

[below middle] A small subsidence sinkhole in karst of class kII, recently re-activated during a rain storm.

[left] Tall limestone towers are the classic features of karst of class kV developed in wet tropical terrains.

[top right] Doline karst of class kIII is a terrain of sinkholes that evolved to leave a net of low polygonal ridges.

[middle right] Conical hills with a variable degree of rounding are typical of mature tropical karst; all the drainage sinks into the floors of dolines between the hills.

[bottom right] A polje is a large, flat-floored valley in mature karst; its drainage is entirely underground and it is normally flooded in the winter or wet season; its margins typically have many solution cavities.

Travertine, or tufa, is a strong calcite crust formed in some karst streams; it may overlie weak sediment and can then be confused with rockhead if not properly assessed.

[left] Upward migration of a cave passage by the progressive, but slow, stoping failure of thin beds of limestone in the roof; the greater threat to structures on the surface is normally from imposed load on a rock roof with a thickness that is significantly less than the width of a cave at shallow depth.

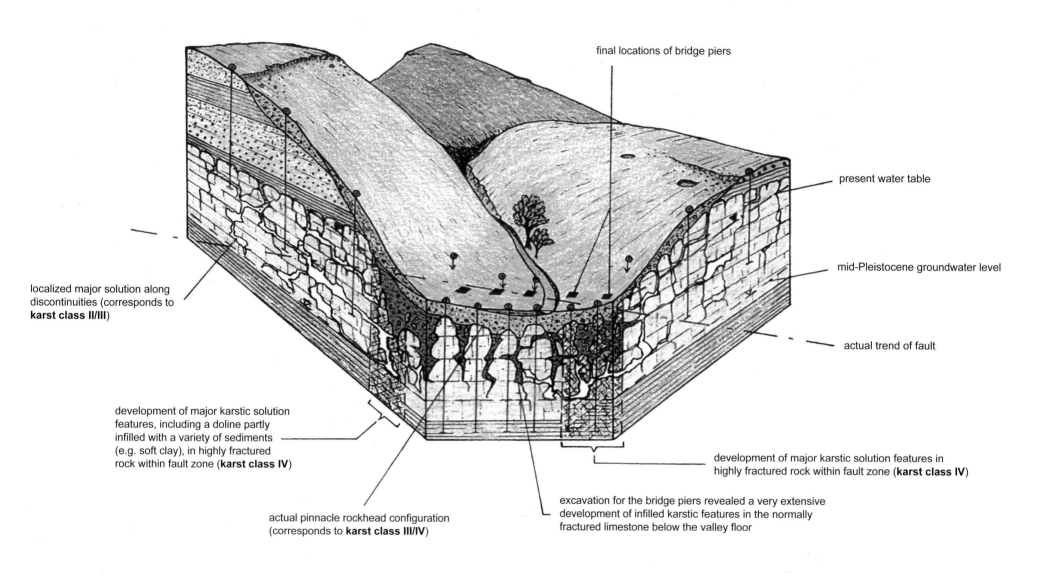

final locations of bridge piers

present water table

mid-Pleistocene groundwater level

actual trend of fault

localized major solution along discontinuities (corresponds to **karst class II/III**)

development of major karstic solution features, including a doline partly infilled with a variety of sediments (e.g. soft clay), in highly fractured rock within fault zone (**karst class IV**)

development of major karstic solution features in highly fractured rock within fault zone (**karst class IV**)

actual pinnacle rockhead configuration (corresponds to **karst class III/IV**)

excavation for the bridge piers revealed a very extensive development of infilled karstic features in the normally fractured limestone below the valley floor

*Figure 5.6 Foundation conditions in limestone bedrock for a river crossing as found during construction.*

## *Karst conditions in limestone foundations for a river crossing as found during construction (Figure 5.6)*

Figure 5.6 shows the actual conditions found during construction and should be compared with Figure 5.5, which shows the conditions anticipated from the ground investigation. Table 5.6.1 gives some types of karst landscapes.

Given the difficulties in obtaining a good subsurface picture of any karst situation, Table 5.6.2 is helpful in diagnosing the class of karst in the area under investigation. The initial classification of karst at a site is based on geomorphological and geological observations/mapping in the area, the published literature and discussions with local authorities and residents, engineers, geologists and caving societies. This preliminary fieldwork and desk study should be carried out before the main ground investigation; see Figures 4.2–4.6 for a guide to the subsequent staged investigation of the karst in a project. Realistic ground investigations of karst areas require geologists and geomorphologists with special training and/or experience.

Many boreholes are needed in the ground investigation to map the pinnacled rockhead and buried sink holes and many rock probes are required to prove solid rock without caves. The local and site history are the best guides to cave and sink hole hazards. Boundaries between limestone and slate and the line of faults may have concentrated the development of sink holes and caves. Deep probes should prove bedrock to depths at least 0.7 times the probable cave width; the site may require inclined borings to prove that the pinnacles are sound.

Subsidence sink holes are believed to account for >90% of ground collapses in limestone. They form in soil cover, above cavernous rock, as a result of the down-washing (also known as suffusion or ravelling) of soil into bedrock fissures. Sinkholes may be 1–100 m across. In field investigations, their potential location is unpredictable, but they are most common in soils 2–15 m thick. Extensive drilling and probing may help to find caves or cavities lying below

potential sink hole locations. In non-cohesive sandy soils, the surface typically subsides slowly. In cohesive clay soils, a cavity forms first at the rockhead, then grows in size until the soil arch fails, causing a sudden dropout collapse of the surface.

### FOUNDATIONS ON LIMESTONE

The most important single measure to prevent new subsidence sink holes forming during and after construction is to control the drainage over or into soils above limestone.

- *Driven piles.* These may lose integrity where they bear on rock over a cave or they may be deflected as a result of meeting a pinnacled rockhead. They may also be founded on loose blocks (floaters) incapable of taking the full pile load, or on unstable pinnacles within the soil.

- *Ground beams.* These may be aligned or extended to bear on rock pinnacles that have been proved sound; stone-filled pads stiffened with geogrids may act in the same way and avoid loading the intervening soil. Stiff grout can be injected to strengthen soils over limestone and prevent suffusion into fissures, but the injection of a fluid grout can incur large losses into adjacent caves before sealing the karstic fissures.

- *Strip or raft foundations.* These can be designed to span any small failures that may develop during or after construction.

Fissures are opened by dissolution until they take all the surface drainage underground. They can then evolve into an infinite diversity of cave passages and chambers. In many limestones, most caves are <10 m across, but some tropical areas have cave chambers >100 m wide. Bedding planes, changes in rock types and discontinuities influence the shape of most cave passages. Cave locations within limestone are unpredictable and commonly have no surface indication. Although isolated cavities cannot exist, their entrances may lie hidden beneath soil cover or may be only small tortuous fissures. Cave roof collapse is only likely under structural loading where the

limestone cover thickness is less than about 0.7 times the cave width. Small individual cavities may allow punching failure and can threaten the integrity of individual piles or column bases. Statistically, most caves are deep enough to have no direct influence on surface engineering.

Table 5.6.1 Types of karst landscape

| Broad types of karst terrain are primarily influenced by climate; these below are in sequence from cold mountains to wet tropics, but local variations are numerous and complicated, and scales vary enormously. Ground conditions for engineering works are generally more difficult in the mature karsts developed in the warmer environments. |
|---|
| *Glaciokarst*: with extensive bare rock and limestone pavements: easily assessed for engineering works except where cover of glacial till masks open fissures and potential sinkhole sites. |
| *Fluviokarst*: dominated by dry valleys formed in periglacial conditions but dry since because water now sinks into unfrozen ground; active caves may be at shallow depth. |
| *Doline karst*: with landscape dominated by sinkholes (also known as dolines) between polygonal net of ridges (so also known as polygonal karst); subsidence potential on sinkhole floors. |
| *Cone karst*: numerous small stream sinks in large sinkholes between conical or hemispherical hills; large caves are likely to exist beneath the sinkhole floors and beneath the conical hills; also known as kegelkarst, cockpit karst and fengcong. |
| *Tower karst*: with isolated steep-sided towers scattered across a karst plain, where limestone is beneath alluvial cover; new sinkholes are common in the alluviated plains when local drainage is disturbed; also known as turmkarst and fenglin. |
| *Pinnacle karst*: dominated by sharp-edged limestone pinnacles; large pinnacles make very inhospitable terrain for any development; small pinnacles may be buried to create very uneven pinnacled rockhead. |

Table 5.6.2 An engineering classification of karst (work this table with the text examples given in Figures 4.2–4.6).

| Karst class | Location | Sink holes | Rockhead | Fissuring | Caves | Ground investigation | Foundations |
|---|---|---|---|---|---|---|---|
| Class kI, juvenile | Only in deserts and periglacial zones, or on impure carbonates | Rare, NSH [a] <0.001 | Almost uniform; minor fissures | Minimal; low secondary permeability | Rare and small; some isolated relict features | Conventional | Conventional, with consideration given to provision in the contract for the observational method (see Figure 4.6) |
| Class kII, youthful | The minimum in temperate regions | Small suffusion or dropout sink holes; open stream sinks; NSH 0.001–0.05 | Many small fissures | Widespread in the few metres nearest to the surface | Many small caves, most <3 m across | Mainly conventional; probe rock to 3 m, check fissures in rockhead | Grout open fissures; control drainage |
| Class kIII, mature | Common in temperate regions; the minimum in the wet tropics | Many suffusion and dropout sink holes; large dissolution sink holes; small collapse and buried sink holes; NSH 0.05–1.0 | Extensive fissuring; relief of <5 m; loose blocks in cover soil | Extensive secondary opening of most fissures | Many <5 m across at multiple levels | Probe to rockhead; probe rock to 4 m; microgravity survey | Rafts or ground beams; consider geogrids; driven piles to rockhead; control drainage |
| Class kIV, complex | Localized in temperate regions; normal in tropical regions | Many large dissolution sink holes; numerous subsidence sink holes; scattered collapse and buried sink holes; NSH 0.5–2.0 | Pinnacled; relief of 5–20 m; loose pillars | Extensive large dissolution openings on and away from major fissures | Many >5 m across at multiple levels | Probe to rockhead; probe rock to 5 m with splayed probes; microgravity survey | Bored piles to rockhead or cap grouting at rockhead; control drainage and abstraction |
| Class kV, extreme | Only in wet tropics | Very large sink holes of all types; remnant arches; soil compaction in buried sink holes; NSH >1 | Tall pinnacles; relief >20 m; loose pillars undercut between deep soil fissures | Abundant and very complex dissolution cavities | Numerous complex three-dimensional cave systems with galleries and chambers >15 m across | Make individual ground investigation for every pile site | Reinforce soils with geogrid; load on proved pinnacles or use deep bored piles; control all drainage and control abstraction |

[a] NSH = approximate rate of formation of new sink holes per km2 per year.

Catastrophic sinkhole collapse in gypsum karst, at a site undermined by a breccia pipe forming in the gypsum beneath 15 metres of soil cover.

Road cutting with breached caves that have been closed by masonry except for access openings; the road was built on an impermeable base to avoid disturbing ground drainage.

[above right] Pipeline that has been left suspended over a new sinkhole, which formed in the soil cover where the drainage had been disturbed by excavation of the pipeline trench.

[below right] Viaduct pier that was relocated after an unpredicted cave (now beneath the grassy slope) was exposed only during excavation for the footing because it had been missed by all the investigation boreholes in karst of class kII.

Road destroyed by a new sinkhole where the underlying soil was lost into karst limestone of class kIII when inflow and suffusion were increased by water-table decline due to pumped drainage of a nearby quarry.

A new subsidence sinkhole in soil that overlies chalk, formed immediately after a water main had burst nearby.

Buildings and infrastructure on the alluvial flats between the towers in karst of class kV may require foundations down to stable underlying bedrock or risk failure by development of new subsidence sinkholes within the alluvium.

The casing of a bored pile that was constructed through a cave to reach footing on intact limestone beneath.

Houses on a black steel frame that can be jacked up, on salt karst with slow subsidence exacerbated by brine pumping.

A building placed on a reinforced concrete slab that is supported entirely on adjacent pinnacles of strong limestone in karst of class kIV, spanning a void that was filled with clay and is now exposed in the hillside.

A rare benefit of mature cavernous karst in strong limestone, where a railway could be constructed inside a large cave in order to pass right through a high limestone ridge.

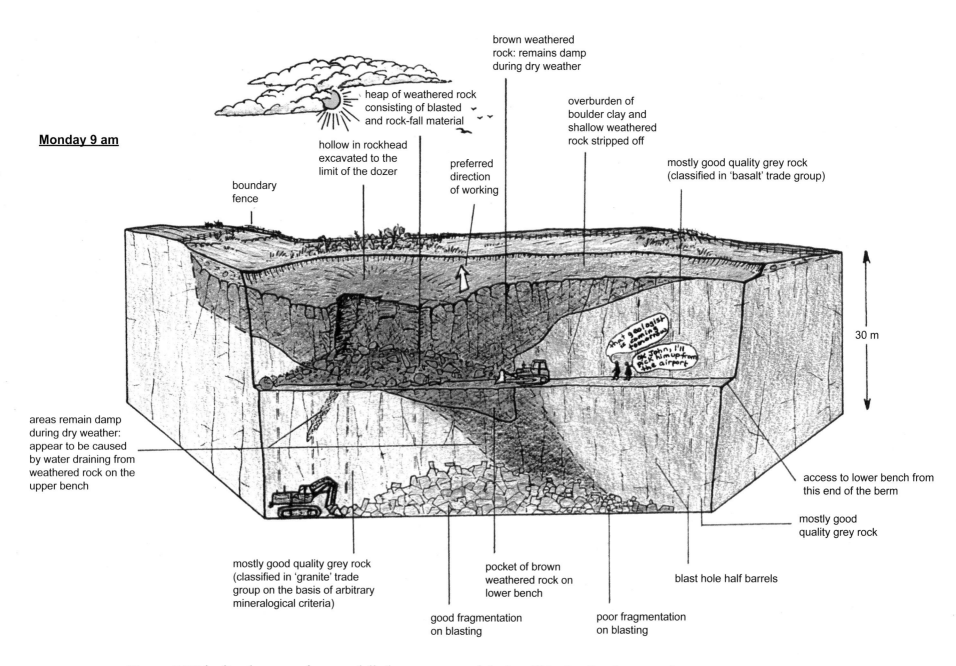

*Figure 5.7 Idealized quarry face modelled on a quartz dolerite sill in Scotland as seen by a quarry manager.*

## Idealized quarry face modelled on a quarry in a quartz dolerite sill in Scotland as seen by a quarry manager (Figure 5.7)

Figure 5.7 and its companion Figure 5.8 are presented to illustrate the differences that may arise between engineers and geologists viewing the same situation. We believe that these differences can stem (at least in part) from their initial outlook, training, experience and their work description. As a broad generalization, engineers tend to be more precise, specific, analytical and quantitative, whereas geologists tend to be more empirical, indirect, rule-of-thumb, qualitative and intuitive (Fookes, 1997).

Figure 5.7 is a diagram of a quarry in Scotland and was made after many years of working in quarries and many discussions with quarry managers. Figure 5.8 is based on a visiting geologist's field notebook compiled while working in the Scottish quarry. The figures are self-explanatory. They do not claim to represent the views of quarry managers or engineering geologists in general.

### Quarries

The worldwide quarrying industry for construction materials is huge. In Britain, the demand for stone is around 200 million tonnes per year. 'Quarry' is used here to mean an excavation in rock and the term 'pit', meaning 'open pit', is used for excavations in superficial and engineering soils, such as a sand pit. Large open metal mines (e.g. Bingham Canyon, USA) are also called pits (Smith and Collis, 2001). The term 'borrow materials' is used for materials won from the ground for civil engineering construction purposes and include rock or soil suitable, with or without processing, for fills, pipeline padding, embankments, breakwaters, roads or concrete and so on.

Aggregate is the term for stone materials used for concrete and road pavements. Sources include superficial or alluvial deposits (mainly river or glaciofluvial sands and gravels) and strong rock (that requires blasting and crushing). Screening of the as-dug or as-crushed material is typically required to sort the aggregate into standard sizes. The suitability of aggregate

Table 5.7.1 Standard aggregate tests.

| Aggregate property | Test procedure (detailed in BS EN 932-5) | Range of value | | Requirements for good roadstone |
|---|---|---|---|---|
| | | *Very good* | *Poor* | |
| Aggregate impact value (AIV) | Percentage fines lost by hammering on standard rig | 5 | 35 | <20 |
| Aggregate abrasion value (AAV) | Percentage loss by abrasion on standard test | 1 | 25 | <10 |
| Polishing stone value (PSV) | Frictional drag recorded on pendulum swing | 70 | 30 | >60 |
| Aggregate crushing value (ACV) | Percentage fines lost by uniform load crushing on standard rig | 5 | 35 | – |
| 10% fines value (10% FV) | Load on standard ACV test rig to give 10% fines loss | 400 | 20 | >100 |
| Flakiness index | Weight percentage particles with minimum thickness <60% mean | 20 | 70 | <40 |
| Water absorption (CBR) | Weight percentage increase after immersion in water for 24 hours | 0.2 | 10 | <2 |
| Frost heave (mm) | Heave of air-cooled column of sample standing in water | <12 | 18 | <12 |
| California bearing ratio | Resistance to plunger penetration, compared with standard | 100 | 60 | >90 |

is usually defined by local standards or codes. Project-specific codes can be developed using experience from similar geo-environments. Careful testing and quality control is essential for successful aggregate production (Table 5.7.1).

The characteristics of rock won from quarries for aggregate depend on the rock type, which should be strong and fresh (i.e. unweathered) and not be heavily tectonized (i.e. not full of fractures). It should also be chemically unreactive with the cement in concrete (see Figure 5.8) and should have good adhesion with bitumen (typically basic rocks) and resistance to polishing if used as a road-wearing course. The shape of the rock depends in part on its characteristics (e.g. foliated metamorphic rocks do not produce cubic shapes, but granite does) and, in part, on the crushing technique (see Table 5.8.1 for natural rock characteristics).

The properties (e.g. shape) of alluvial sand and gravel depend on the rocks from which they were derived. Gravel is typically composed of several rock types, but sands are composed mainly of quartz or carbonates. After erosion from the parent rock and subsequent natural transportation by winds, rivers or coastal systems, weathered or otherwise weak fragments tend to be selectively worn away. As a result, these

aggregates may be stronger than those produced by crushing the quarried weathered parent material. Some rounding of the particles will occur during transport. Where the natural transport distances are short, the rounding of particles may be less (e.g. in glaciofluvial environments, glaciers or alluvial fans) and impurities (e.g. coal or chalk fragments) may remain in the deposit and reduce the quality of the prospective source. Clay is preferentially carried away by flowing water (e.g. rivers or wet screening) and is therefore typically fairly minor in fine and coarse gravel aggregates. Tables 5.8.1 and 5.8.2 in the next text give the characteristics of the various rock types and potential aggregate and building stone materials, together with the likely sources of borrow materials, in various world geo-environments.

### Aggregates in construction

Aggregates used in construction typically consist of coarse (gravel or crushed rock forming about 50% of the unit weight) and fine (sand about 25% of the unit weight) aggregates, together with cement and water, and additives that improve performance and workability (about 25%). Good quality aggregate for concrete has the following properties.

- The unconfined compressive strength of the aggregate is equal to or greater than that of the cement matrix (i.e. the bonding material) and is therefore typically >80 MPa.

- It is free from deleterious (unsound) materials – that is, materials prone to volume changes or chemical reactions (e.g. natural evaporite salts can initiate rusting of reinforcing bars or can react with cement to cause volume changes).

- It is clean and free from organic impurities – the particles should not be coated by clay or dust and should have low absorption properties.

- It should have a good natural or crushed shape to establish good interlocking between particles and should ideally not be too flaky or elongated as these factors introduce difficulties in workability and strength.

Road aggregates – for example, those for moderately to heavily trafficked conditions with flexible pavements – commonly consist of four layers (i.e. they follow a California bearing ratio (CBR) design).

- *Sub-base.* The unbound drainage layer at the bottom of the road. This derives its strength from the interlocking and compaction of well-graded, natural or processed aggregates (i.e. not single-sized) and is placed on the formation (grade in the USA) level, which is the prepared top of the natural ground or embankment.

- *Base course.* This is the main load-bearing layer of the road. Aggregate is typically processed and needs to have a good strength. It should not be porous (to prevent changes on wetting and drying) and should be non-plastic or of low plasticity (PI < 6) with a reasonable shape and grading. Some road bases are bound by cement, lime or bitumen, others simply by good mechanical interlocking of the particles and compaction.

- *Road base.* This distributes the traffic load from the wearing course onto the base course and provides a highly specified surface on which the wearing course is laid. It is usually bitumen-bound and has very good quality crushed aggregate of high strength and a close tolerance on grading and shape.

- *Wearing course.* This surface layer is commonly bound with bitumen. The highest quality aggregates are required, with high strengths and the ability to resist polishing and abrasion as well as the ability to bond with bitumen. Alternatively, wearing courses and road bases can be designed as non-flexible and can be made of concrete.

The specifications of material for roads are increasingly onerous, from the sub-base upwards, notably on the particle shape and grading curve (see Table 5.7.1). All aggregates must be strong and clean with limits on the soluble salts and unsound materials. Most roadstone in the upper layers is therefore produced from crushed rock because natural gravels tend to be too rounded, polished and with more impurities. The specifications for roads in Britain are given in the *Specifications for Highways Work* (Highways Agency, 2007, or latest edition) and various test procedures are given by, for example, Smith and Collis (2001).

[top left] The upper bench of an aggregate quarry lies on the top of the strong limestone, and the upper face is an overburden of shales and glacial till that has to be cleared.

[middle left] A limestone quarry has shot holes drilled on its main bench, below an upper face that has exposed a clay-filled cave, a common hazard in limestone quarries.

[bottom left] Pumped drainage is a major cost in quarries extending below the water table, especially in karst limestones.

[top] Multiple benches and a system of haul roads designed in advance are essential in a deep open-pit copper mine.

[above] Large dump-trucks and shovels provide the economies of scale that are necessary in working a thick coal seam.

A deep open mine on a vertical mineral vein, with abandoned faces that are higher than would be acceptable in any modern quarry.

This large segment of wallrock subsided into an abandoned vein mine, which has since been backfilled to prevent wider ground failure.

[above] After blasting within a quarry in strong granite with widely spaced fractures, large blocks are selected for cutting into dimension stone, and all the remaining rock is crushed to produce aggregate.

[right] A quarry in uniform granite goes deep to avoid any weathered material; its sawn blocks are cut and polished for cladding stone.

[top left] A pattern of shot holes, each 15 metres deep, is drilled in preparation for blasting of a single bench in a large open-pit mine.

[middle left] Blasting that utilises delayed firing, so that the outer set of holes are fired some milliseconds before the inner set of holes.

[bottom left] An efficient blast in a limestone quarry produces debris of uniform size designed to fit the gape of the primary crusher.

[above] An old limestone mine had 80% extraction, but its narrow pillars safely support a rock cover less than 10 metres thick.

[right] A granite quarry worked for dimension stone has an upper face that was cut by line drilling, and a lower face cut by water jet.

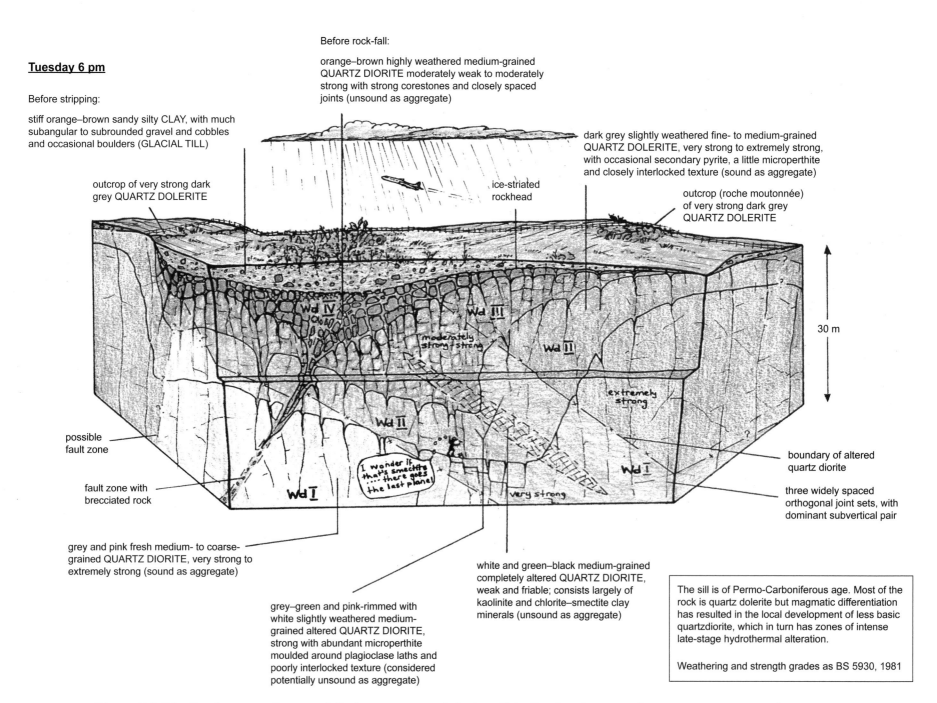

**Tuesday 6 pm**

Before stripping:

stiff orange–brown sandy silty CLAY, with much subangular to subrounded gravel and cobbles and occasional boulders (GLACIAL TILL)

Before rock-fall:

orange–brown highly weathered medium-grained QUARTZ DIORITE moderately weak to moderately strong with strong corestones and closely spaced joints (unsound as aggregate)

dark grey slightly weathered fine- to medium-grained QUARTZ DOLERITE, very strong to extremely strong, with occasional secondary pyrite, a little microperthite and closely interlocked texture (sound as aggregate)

outcrop of very strong dark grey QUARTZ DOLERITE

ice-striated rockhead

outcrop (roche moutonnée) of very strong dark grey QUARTZ DOLERITE

30 m

Wd IV

Wd III

Wd II

moderately strong/strong

extremely strong

possible fault zone

Wd II

*I wonder if that's smectite .... there goes the last plane!*

very strong

Wd I

boundary of altered quartz diorite

fault zone with brecciated rock

Wd I

three widely spaced orthogonal joint sets, with dominant subvertical pair

grey and pink fresh medium- to coarse-grained QUARTZ DIORITE, very strong to extremely strong (sound as aggregate)

grey–green and pink-rimmed with white slightly weathered medium-grained altered QUARTZ DIORITE, strong with abundant microperthite moulded around plagioclase laths and poorly interlocked texture (considered potentially unsound as aggregate)

white and green–black medium-grained completely altered QUARTZ DIORITE, weak and friable; consists largely of kaolinite and chlorite–smectite clay minerals (unsound as aggregate)

The sill is of Permo-Carboniferous age. Most of the rock is quartz dolerite but magmatic differentiation has resulted in the local development of less basic quartzdiorite, which in turn has zones of intense late-stage hydrothermal alteration.

Weathering and strength grades as BS 5930, 1981

*Figure 5.8 Idealized quarry face modelled on a quartz dolerite sill in Scotland as seen by an engineering geologist.*

## Idealized quarry face modelled on a quarry in a quartz dolerite sill in Scotland as seen by an engineering geologist (Figure 5.8)

Figure 5.8 should be worked with Figure 5.7. Again, the figure is self-explanatory and illustrates the view of the same quarry, but now as seen by a visiting engineering geologist. The skill in the management of the quarry layout and blasting arrangements to a large extent dictates the as-blasted size of the rock, which has to be suitable for the quarry's processing flow path and the size and type of crushing and screening equipment. Good quarry management leads to the cost-effective manufacture of standard aggregate sizes. The type of rock and the way it is crushed dictate its aggregate shape.

Table 5.8.1 summarizes the probable construction characteristics of the principal rock types, tabulated using the old trade group classification BS 812 Part 1 (1975), which lists together rocks of similar engineering behaviour when in their fresh (unweathered) state.

The *alkali silica reaction*, the most common form of alkali aggregate reaction, can be caused by some rock types in which an expansive reaction leads to the cracking of concrete. This is a result of the reaction of alkaline solutions in the cement matrix with siliceous aggregates to form an alkali silica gel, which swells when it imbibes water. It can occur (depending on the local rock type) more or less anywhere in the world. Britain has had expensive alkali silica reaction problems with some of its marine-dredged aggregates. Other forms of alkali aggregate reaction (e.g. the alkali carbonate reaction) are less common.

Dimension stone, including that which is polished for use in floors or cladding, is worked mainly without blasting in quarries set up for wire-cutting, sawing or high-pressure water jets. Dimension stone is also quarried in underground mines (Smith, 1999).

Table 5.8.2 is based on the world's geo-environments given in earlier sections. It summarizes the typical locations where borrow materials worked by pits, quarries or draglines under water can be found.

Table 5.8.1 Rock types and their potential aggregate and building stone characteristics.

| Group | Including | Common characteristics | Roadstone quality | Concrete quality | Building stone quality |
|---|---|---|---|---|---|
| Artificial | Crushed brick, slag, calcined bauxite | Man-made | Varies, generally poor except in lower parts of pavement | Varies, some good | Typically poor, not suitable |
| Basalt | Dolerite, andesite, spilite, epidiorite | Strong, fine-grained, basic igneous | Typically good | Typically good; beware andesite alkali reactivity | Typically good |
| Flint | Chert | Fine-grained silica, mostly as gravel; may be brittle | Generally poor | Generally good; beware alkali reactivity | Generally good |
| Gabbro | Diorite, basic gneiss, amphibolite, norite, serpentinite | Strong, coarse-grained, basic igneous and metamorphic | Generally good | Generally good | Generally good |
| Granite | Gneiss, granodiorite, pegmatite, syenite | Strong, coarse-grained, acid igneous and metamorphic | Typically good | Typically good | Good; large size commonly available |
| Gritstone | Greywacke, grit, arkose, sandstone, lithified tuff | Stronger, well-cemented sandstones | Typically good | Typically good | Good; large size commonly available |
| Hornfels | Contact altered rock of all kinds, except marble | Strong, fine-grained, acid igneous rocks | Typically good | Generally good | Generally good |
| Limestone | Marble, dolomite | Stronger limestones and dolomites | Good, except for wearing course as it polishes | Typically good | Generally good; large size commonly available |
| Porphyry | Aplite, dacite, felsite, rhyolite, trachyte | Strong, fine-grained | Generally good | Generally good; beware alkali activity | Generally good |
| Quartzite | Ganister, quartzitic sandstones, recrystallized quartzite | Strong, metamorphosed sandstone | Generally good; beware flaky types | Generally good; beware alkali activity | Generally good; large size available |
| Schist | Slate, phyllite, all severely sheared rocks | Flaky, sheared or cleaved metamorphic rocks | Poor, often too flaky | Varies, often too flaky | Varies, often too flaky |
| Others | All weak sedimentary rocks (including chalk) | Easily broken down or weathered | Only lower parts of road (but commonly acceptable for gravel roads) | Generally too poor | Varies, often poor |

Table 5.8.2 World morphoclimatic environments (Figures 1.1 and 1.2) and potential borrow sources.

| Geo-environment | Comment | Landforms and materials |
|---|---|---|
| Glacial (Figure 3.1) | Extremely variable, often unsorted; typically needs processing with washing | *Till*: unsorted material, unlikely to be potential aggregate source unless matrix is sandy; possible fill<br>*Hummocky drift*: unsorted, low aggregate potential<br>*Drumlins*: may contain an upstanding core of solid rock: possible fill<br>*Glacial outwash plains (glaciofluvial)*: melt water deposits, relatively good aggregate sources<br>*Kames and eskers*: ice-contact deposits, relatively good aggregate sources |
| Periglacial (Figure 3.2) | Periglacial conditions produce little sand and gravel (except locally in river valleys), but may provide cobbles and boulders for crushing | *Screes and talus cones*: provide a ready source of coarse aggregate, commonly poor quality due to weathering<br>*Rock fields and boulder streams*: may provide rock suitable for crushing to aggregate; beware environmental damage |
| Fluvial (any environment) | Alluvium commonly a source of large volumes of dredged sand and gravel; relict Quaternary sources may lie beneath contemporary flood plains; generally needs processing | *River terraces*: useful sources of aggregate, but highly variable as a result of former channel migration (evulsion)<br>*Alluvial fans*: useful coarser sand and gravel deposits at the proximal (upstream) end; commonly poorly sorted and interbedded with silts and clays |
| Hot drylands (Figure 3.6) | Dominant upward leaching of salts causes aggressive ground conditions and contamination of materials; larger particles may be weathered; coarser aggregate sources may be difficult to find, except in mountainous regions | *Coastal sabkhas, inland salinas*: surface crusts (e.g. sodium chloride, calcium and calcium–magnesium carbonates, calcium sulphate) commonly severely contaminate sands and gravels; high quality, strong control essential; avoid if possible<br>*Coastal deposits*: carbonate sands may be acceptable fine aggregates; generally require processing (e.g. screening and washing)<br>*Sand sheets and dunes*: rounded single-sized particles can be used as concreting sands and for embankments with careful mix design; possible contamination by wind-blown salts<br>*Alluvial fans*: source of poorly sorted sands and gravels around margins of uplands; may be salt-free as a result of downward leaching during periodic floods<br>*Ephemeral (occasional flow) rivers*: limited amounts of sands and gravels, especially around margins of uplands; ancient buried river deposits may exist under sands |
| Savanna (Figure 3.7) | River and alluvial fan deposits, plus extensive duricrusts (e.g. laterite, calcrete, silcrete), can generally provide acceptable materials; quarries in strong rock areas generally acceptable, provided not too deeply weathered | *Laterite and calcrete (and other)*: strong case-hardened material may be suitable on crushing for roadstone and can be used for concrete if clean, strong and not friable<br>*Laterite and latosols (lateritic soils)*: may make excellent gravel roads and haul roads on wetting and rolling and may stabilize fills |
| Hot wetlands (Figure 3.8) | Deep tropical weathering results in limited coarse, strong materials; bedrock quarries for crushed aggregates need careful excavation of uncontaminated fresh rock; rivers may provide a variety of borrow materials | *Ancient strong duricrusts (e.g. ferricrete, silcrete)*: may be useful borrow materials which require crushing; beware enviromental damage<br>*Volcanic deposits*: may produce alluvium with suitable sands and gravels, although fragments may contain macro-voids |

Armour stone of 5-tonne blocks of strong gneiss, economically carried 700 km to site by sea.

Limestone rip-rap prevents erosion on a reservoir, but is expensively carried 90 km to site by road.

Screening plant and stockpiles at a pit producing aggregate from alluvial gravel.
[below right] Screened and crushed, rounded gravel of mixed-rock-type at the same site.

A small quarry works high-quality roadstone aggregate from steeply dipping greywacke. The site is constrained by a river on the far side, by the road in the foreground, and by adjacent slates left and right, with the crushing plant and stockpiles on slate to the right.

[right] The primary cone crusher at the same roadstone quarry, with its boom-mounted hammer to reduce any oversize quarry-run blocks.

Water-cooled jig with multiple saw blades cutting slabs of limestone.

Concrete with deep cracks due to alkali-silica reactivity in marine aggregate, in the wall and columns of a multi-storey car park.

Chemical attack on poor concrete, and subsequent corrosion of its steel reinforcement bars, at the inter-tidal level of a marine structure.

Dredged marine shells used as road aggregate; not widely acceptable as the calcite shells polish too easily and offer little skid-resistance.

A concrete beam supporting a freeway, with its reinforcing bars exposed by spalling after corrosion of the steel in a concrete mixed with salty water.

[above] Poor quarry after-use; this bungalow was destroyed by methane that filled it and exploded after flowing up-dip through a porous sandstone from a nearby quarry, which had been filled with rubbish and then sealed beneath a clay cap.

[right] Good quarry after-use; a face that ended on a steep bed of mudstone is preserved with a pair of dinosaur tracks, among others also of world class on the same bed.

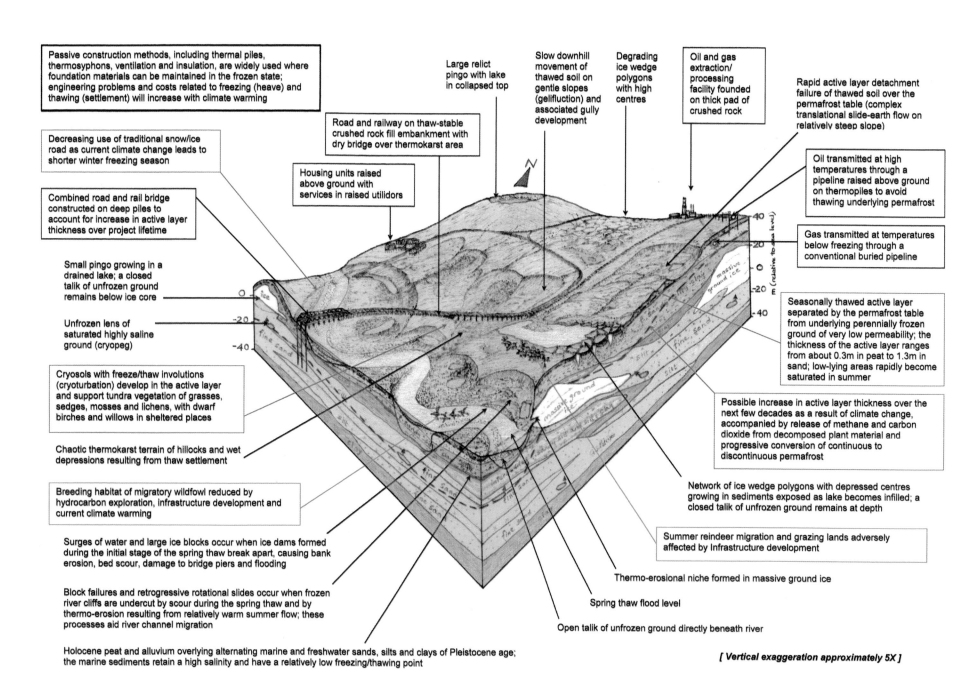

Passive construction methods, including thermal piles, thermosyphons, ventilation and insulation, are widely used where foundation materials can be maintained in the frozen state; engineering problems and costs related to freezing (heave) and thawing (settlement) will increase with climate warming

Large relict pingo with lake in collapsed top

Slow downhill movement of thawed soil on gentle slopes (gelifluction) and associated gully development

Degrading ice wedge polygons with high centres

Oil and gas extraction/ processing facility founded on thick pad of crushed rock

Rapid active layer detachment failure of thawed soil over the permafrost table (complex translational slide-earth flow on relatively steep slope)

Road and railway on thaw-stable crushed rock fill embankment with dry bridge over thermokarst area

Oil transmitted at high temperatures through a pipeline raised above ground on thermopiles to avoid thawing underlying permafrost

Decreasing use of traditional snow/ice road as current climate change leads to shorter winter freezing season

Housing units raised above ground with services in raised utilidors

Gas transmitted at temperatures below freezing through a conventional buried pipeline

Combined road and rail bridge constructed on deep piles to account for increase in active layer thickness over project lifetime

Seasonally thawed active layer separated by the permafrost table from underlying perennially frozen ground of very low permeability; the thickness of the active layer ranges from about 0.3m in peat to 1.3m in sand; low-lying areas rapidly become saturated in summer

Small pingo growing in a drained lake; a closed talik of unfrozen ground remains below ice core

Unfrozen lens of saturated highly saline ground (cryopeg)

Cryosols with freeze/thaw involutions (cryoturbation) develop in the active layer and support tundra vegetation of grasses, sedges, mosses and lichens, with dwarf birches and willows in sheltered places

Possible increase in active layer thickness over the next few decades as a result of climate change, accompanied by release of methane and carbon dioxide from decomposed plant material and progressive conversion of continuous to discontinuous permafrost

Chaotic thermokarst terrain of hillocks and wet depressions resulting from thaw settlement

Network of ice wedge polygons with depressed centres growing in sediments exposed as lake becomes infilled; a closed talik of unfrozen ground remains at depth

Breeding habitat of migratory wildfowl reduced by hydrocarbon exploration, infrastructure development and current climate warming

Summer reindeer migration and grazing lands adversely affected by infrastructure development

Surges of water and large ice blocks occur when ice dams formed during the initial stage of the spring thaw break apart, causing bank erosion, bed scour, damage to bridge piers and flooding

Thermo-erosional niche formed in massive ground ice

Block failures and retrogressive rotational slides occur when frozen river cliffs are undercut by scour during the spring thaw and by thermo-erosion resulting from relatively warm summer flow; these processes aid river channel migration

Spring thaw flood level

Open talik of unfrozen ground directly beneath river

Holocene peat and alluvium overlying alternating marine and freshwater sands, silts and clays of Pleistocene age; the marine sediments retain a high salinity and have a relatively low freezing/thawing point

*[ Vertical exaggeration approximately 5X ]*

*Figure 5.9 Example of infrastructure development in a sensitive environment: the continuous permafrost zone.*

## Example of environmental impact and infrastructure development in a sensitive environment: the continuous permafrost zone (Figure 5.9)

Figure 5.9 shows the characteristic periglacial features in the zone of continuous permafrost encountered in the Arctic regions of Alaska, Canada and Siberia (see Figures 1.1, 1.2, 2.2, 3.2 and 3.5; also Walker, 2005). The permafrost in these areas extends to depths of 100–1000 m, averaging about 400 m. An upper (active) layer thaws during summer to depths of about 0.3 m in peaty soils and 1.3 m or more in sandy soils. Perennially frozen ground below the subsurface permafrost table has a very low permeability and typically contains ice wedges and lenses of massive ground ice, including those that expand upwards to form pingos. Networks of actively growing ice-wedge polygons have depressed centres that usually contain water in summer, whereas networks of old degrading polygons have high centres bounded by wet linear depressions.

Temperature-change erosional processes interact to form a distinctive landscape surface of cryoplanation (i.e. dominated by frost action) with intervening thermokarst terrain of small hills and boggy depressions caused by melting of the underlying ground ice. Cryogenic (freeze–thaw) failures are common and include slow gelifluction (cold-climate solifluction) movement of soil in the active layer on slopes as gentle as 1–2°. Rapid active layer detachment slides and flows of saturated soil masses occur on steeper slopes.

Figure 5.9 is based mainly on the Yamal Peninsula in western Siberia, where thermo-erosion and thermokarst processes have been exacerbated in recent decades by exploration for oil and gas and the associated infrastructure development (Sidorchuk and Grigor'ev, 1998). The mean annual air temperature in the southern half of the peninsula is −8°C, ranging from −24°C in the coldest month to +6°C in the warmest month. The mean annual ground temperature is −6°C and the permafrost is about 150 m thick.

A gently undulating terraced landscape with meandering rivers and numerous lakes has developed in marine and freshwater sands, silts and clays of Pleistocene age, overlain by Holocene peat and alluvium. Saline water derived from the marine sediments accumulates where the permafrost table is depressed and aids the formation of unfrozen ground (taliks). Surface depressions become waterlogged in summer, favouring the formation of cryosols (previously frozen soils within 1 m of the ground surface) that have gleyed horizons, characterized by a low iron content, below a surface layer of peat. There are also many gullies formed both by natural processes and by poorly controlled run-off from artificial structures. Typical tundra vegetation of grasses, sedges, mosses and lichens, with dwarf willows and other shrubs in more sheltered locations, provides breeding grounds for migratory wildfowl and summer pastures for the reindeer herds managed by nomadic people. It is difficult, but not impossible, to reconcile traditional practices with modern industrial development.

### ENGINEERING PROBLEMS

Most engineering problems in periglacial environments are related to freeze–thaw processes, particularly in poorly drained fine-grained sediments. Permanently frozen soils provide a good foundation for structures, but the soil strength within the active layer is substantially reduced. Many roads, railways and airfields undergo some degree of thaw settlement in summer and frost-heave in winter; piles are vulnerable to frost-jacking and weakening of the adfreeze bond.

- Active construction methods, where permafrost degradation cannot be prevented, include thawing and compacting or replacing unfavourable thaw-unstable formation/subgrade materials (e.g. fine soils and sediments) with thaw-stable (e.g. coarse cohesionless soils and sediments).
- Passive construction methods, widely used where foundation soils can be kept frozen (permafrost conservation), include thermal piles, thermosyphons, ventilation, insulation and the use of crushed rock or natural gravel for thaw-stable embankments and pads.

- Roads and railways are best constructed on 2–3 m of thaw-stable granular fill to contain seasonal thawing within the embankment.
- Open spaces below buildings, raised utilidors and dry railway bridges allow cold air to circulate so that the ground can follow its normal annual temperature cycle.

Many of these design solutions are very expensive to construct and maintain. Suitable fill materials may have to be transported considerable distances from hard rock quarries in the northern Russian Urals and the Alaskan Brooks Range.

### CLIMATE CHANGE

Periglacial environments are very sensitive to climate change. The mean annual air temperatures in the Arctic have risen by about 1.5°C over the last century and an additional rise of about 7°C could occur by 2100, a rate of warming at least twice the current global average (EMERCOM, 2013; Ria Novosti 2013). River ice in the Yamal Peninsula now forms later in the year and breaks up earlier; the thickness of the active layer is currently increasing by 2 cm a year. A further temperature rise of 1.5°C in the Arctic could lead to extensive thawing of the upper permafrost layers, potentially releasing carbon dioxide and methane gases derived from decomposed plant material. Some areas, such as Baffin Island in Canada (Miller *et al.*, 2013), may already be on the threshold of such changes. An increase of about 2°C globally (i.e. at least 4°C in the Arctic) would result in temperatures similar to those during the Last (Eemian) Interglacial Period 120,000 years ago, when the global sea level was several metres higher than at present.

Engineering problems are expected to become more widespread if the predicted global climate change occurs. Thermokarst development will accelerate as temperatures rise and the boundary between the discontinuous and continuous permafrost will probably shift northwards (in the northern hemisphere) and ground conditions will therefore become more unpredictable. Increasingly expensive construction techniques, including substantial retrofitting of existing structures, are likely to be required.

A large pingo rises 45 metres above an alluvial plain.

Thin soil and plant cover in the fragile Arctic environment.

[above] Frost mounds are a widespread consequence of the annual expansion of ground ice in the active layer.

[left] Periglacial rivers are distinguished by their wide plains of gravel deposited by annual meltwater floods.

[below left] A lens of ground ice within alluvial sands, exposed in an experimental tunnel into the permafrost.

In the wall of a tunnel 8 metres below ground level, glaciofluvial sands (pale buff) containing 80% ice (black) that was drag folded beneath an ice sheet.

Annual ice break-up on an Arctic river can temporarily halt crossing traffic, between a winter ice bridge and a summer ferry, and can damage permanent structures.

Thaw lakes form and then expand wherever the vegetation cover is disturbed on a permafrost plain.

Enabling cold winter air to circulate freely beneath warm buildings ensures conservation of the permafrost; [left] house on wooden blocks on a gravel pad; [right] company building elevated on piles founded in ground ice.

Linking houses elevated on piles, utilidors carry inbound water and outbound waste inside insulated tunnels perched on timber piles, to eliminate heat losses into the permafrost.

This house subsided when its own heat loss thawed the ground ice within unstable clay-rich alluvium.

Steel ducts carry cold winter air through a gravel pad to conserve the permafrost beneath a large, warm strorage tank.

Carrying warm oil, the Trans-Alaska Pipeline is supported on trestles with piles into the frozen ground beneath the active layer; each pile has internal ammonia refrigeration with rising gas and sinking liquid that transfers heat from the ground to radiator fins above, and thereby conserves the permafrost.

[left] A sea of mud has been created when fragile vegetation cover was destroyed by uncontrolled construction works on the active layer over permafrost.

[right] A road's gravel bank provides insulation to conserve the permafrost.

# Appendix. Geotechnical problems associated with different types of engineering soils

Appendix. Geotechnical problems associated with different types of engineering soils (adapted and extended from Fookes, 1997a).

| Soil type and related figures | Typical properties and characteristics | Ground investigation and testing problems | Construction and materials problems |
|---|---|---|---|
| **Transported alluvial and 'engineering' soils (all Part 4 figures)** | Strength typically decreases and compressibility increases with increasing fines content | | |
| *Gravels* (Figures 1.6, 1.7, 3.1, 3.2, 3.3, 3.4, 3.6, 3.9, 5.4) | Can be gap-graded with voids only partially filled with fines, which may then migrate if hydraulic gradient is increased; strength high, compressibility low; densities vary widely according to mode of deposition; permeability variable, often very high, depending on grading and packing | Difficult to investigate by boring; difficult to obtain representative samples from below water-table as fines are washed out during drilling and sampling; thin layers of other soils may not be detected; permeability best determined by pumping tests | Generally good foundations, full consolidation occurs during construction; large water flows into excavations; generally good for use as 'granular' fill and coarse aggregates in concrete, but may contain chemical impurities that react with cement minerals (e.g. pyrite); single-size gravels are self-compacting when deposited in water, but fills choked with silt and clay matrix can be difficult to compact when wet and may not lose water readily under gravity drainage |
| *Sands and silts* (Figures 1.6, 1.7, 3.1, 3.2, 3.3, 3.4, 3.6, 3.7, 3.9, 5.4) | Engineering properties usually improve with geological age; densities vary widely according to mode of deposition; loose sands and silts are very susceptible to liquefaction during earthquakes, can develop flow slides and may undergo large settlements when subject to vibration; permeability moderate to high; very erodible, with risk of surface erosion by water and wind, and piping due to internal water flow; possibility of collapse of dry sands on wetting, particularly if weathered; silts and fine sands are susceptible to frost heave | Often weakly bonded or with interlocking grains, although these characteristics are lost on disturbance; very difficult to obtain undisturbed samples as the ground at the base of boreholes is often disturbed by water inflow when drilling below the water-table; SPT values may be low in sand as a result of soil disturbance or high in silt as a result of pore pressure effects; thin interlayers of other soil types may not be detected | Sandy soils generally provide good foundations and consolidate during construction; however, there are risks of base failures in excavations by piping, loss of soil through sheet pile clutches, soil and water inflows into tunnels and high abrasion of tunnelling machines; generally good fill material, but dry single-size sands have poor trafficking characteristics; silty soils may not drain by gravity during wet weather, resulting in poor compaction of fill, poor trafficability (bounce), inadequate subgrade for road and increased fuel consumption of plant; low density loose dumped moist sands and silts may collapse on inundation |
| *Under-consolidated clays* [a] (not yet fully compressed under current stresses, due to rapid deposition or very recent additional loading) (Figures 1.6, 1.7, 5.4) | Typically very soft clays, silts and muds with excess pore pressures (e.g. estuarine deposits); very low undrained strength and high compressibility relative to depth; may have developed a dried surface crust | Access for drilling rigs may be restricted by soft ground and tidal inundation; very difficult to sample as a result of very low strength; need to use piston sampling; good quality vane tests may be useful | Ground surface still settling; excavations for ports and for cut and cover tunnels are likely to be unstable; methods used to avoid large settlements include dig and replace, fill surcharge, geotextiles, wick drains, tyre bales and timber fascine rafts |
| *Normally consolidated clays* [a] (current effective stress is the maximum to which the soil has been subjected) (Figures 2.1, 5.4) | Low undrained strength; high compressibility and secondary compression (creep) that increase with plasticity; tendency of clay-fill to swell or heave increases with plasticity; exposed surfaces are usually over-consolidated by desiccation | Sensitive clays difficult to sample without undue disturbance; need to use thin-walled tube sampling; in situ vane tests allow measurement of loss of strength on remoulding (sensitivity); cone penetrometer testing can locate thin layers of silt and sand; potential sources of fill are assessed using in situ and laboratory tests | Low allowable loading pressures under structures and embankments; large post-construction settlements; base heave and failure in strutted excavations with high strut loads; down-drag on piles; low strength and difficult working conditions for plant during excavation; acceptability of fill depends on water content and plasticity |
| *Over-consolidated clays* [b] (current effective stress is less than previous maximum; the pre-consolidation pressure is commonly a result of overlying sedimentary deposits subsequently removed by erosion over geological time, but sometimes resulting from loading if ice sheets or tectonic compression) (Figures 3.3, 3.7, 3.8, 5.4) | Higher undrained strengths than for normally consolidated clays, but difficult to predict, and much lower compressibility; generally have low, sometimes negative, in situ pore pressures; permeability may be controlled by flow through extensive clay fissures; if effective stresses due to engineering work exceed the pre-consolidation pressure, behaviour reverts to that of normally consolidated clay | Test results may be affected by sampling disturbance, presence of fissures in older clays, etc.; rotary coring gives better samples than cable percussion drilling in heavily over-consolidated clays and is more likely to reveal rupture surfaces of deep-seated rotational slides; SPT tests often useful back-up to laboratory testing; specimens of material described in the field as hard clay or very weak mudstone may slake in water; degree of over-consolidation controls optimum water content for compaction | Mass strength of foundations commonly affected by fissuring. Pre-existing shear surfaces, particularly in highly plastic clays, may control slope stability. High in situ horizontal stresses in heavily over-consolidated clays can cause large horizontal movements during and after excavation and high lateral stresses on buried structures; possibility of long-term swelling when used as fill; lightly over-consolidated clays are more likely to have acceptable properties |

[a] The majority of clays (but not all) are deposited in the sea or modern estuaries; some young clays in the Quaternary were deposited in lakes (lacustrine clays), some in rivers (alluvial clays) and some clays were derived from rock weathering (residual clays).

[b] Mainly ancient bedrock clays in the Mesozoic (Permian, Triassic, Jurassic) and Tertiary, also some Quaternary tills.

| Soil type and related figures | Typical properties and characteristics | Ground investigation and testing problems | Construction and materials problems |
|---|---|---|---|
| **Other transported sedimentary soils** | | | |
| *Taluvium (coarse colluvium)* (all Part 3 figures) | Heterogeneous slope gravity-moved coarse debris; typically originates as permeable talus (scree) or rock avalanche deposits; as voids become infilled with slope-wash, air-fall volcanic ash or loess, pore pressures increase and shear strength is reduced; shallow debris slides, triggered by seasonal rains or earthquakes, are common and often disintegrate into debris flows down-slope | Usually includes all sizes of material up to large boulders; very difficult to drill and difficult to assess 'average' conditions; field mapping combined with trial pits may give a good indication of ground conditions, but seasonal variations in the water-table, pore pressure and shear strength may not be recognized in tropical wet/dry climates | Cut and fill operations may reactivate landslide masses at limiting equilibrium or initiate new movements; talus deposits may develop avalanche or flow slide behaviour when disturbed during excavation |
| *Fine colluvium (including active solifluction deposits)* (all Part 3 figures) | Heterogeneous slope-wash fine debris; shallow debris or earth slides over a bedrock surface are common in thin clayey colluvial soils on upper slopes saturated by prolonged rainfall or snow melt; deep-seated slides in thicker deposits on lower slopes may occasionally extend down into underlying strata; groundwater pressure may be high in landslide toes | Deep-seated rotational slides often have rupture surfaces marked by a thin colluvial clay layer of low residual strength that may be difficult to recognize in borehole samples; properties measured during a ground investigation may differ from those encountered during construction as a result of seasonal changes in water content | Major and minor landslides may be initiated or reactivated by river erosion or excavations in side-long ground; slopes destabilized by toe erosion may take many decades to achieve limiting equilibrium after the cause of erosion has been prevented; haul road surfaces become slippery for traffic in wet weather |
| *Hot desert soils (not including coastal sabkhas)* (Figure 3.6) | Generally granular and uniformly (mainly single-size) graded with little or no clay, but often large amounts of silt; wind-blown or coastal soils have low density; materials deposited by ephemeral flows in wadis or fans are typically poorly sorted (many different sizes) sands and gravels with angular particles; a near-surface water-table in low-lying areas causes precipitation of evaporite salts | Engineering performance generally related to particle size grading, in situ density and Atterberg limits; some in situ fine-grained soils have anomalously high strength because of high suction pressures or weak cement; low-lying sites need careful investigation for possible high salinity groundwater and low bearing capacity salty soils | Soils may be highly erodible once thin protective stone pavements are removed or disturbed and engineering works may need to be protected from sand storms; wind-blown silts or fine sands are liable to collapse on wetting and loading; salty low-lying ground is highly aggressive to structures and road pavements, duricrusts and densely packed boulders in wadis may cause excavation difficulties, but are potential sources of coarse aggregates; fine aggregates can be in short supply and/or contaminated by salts |
| *Glacial soils* (Figures 3.1, 3.2, 5.3, 5.4) | Often heterogeneous, both horizontally and vertically; grading curve may be almost a straight line over a wide range of particle sizes; alluvial lenses of laminated sand, silt and clay give complex groundwater conditions, including artesian; density and strength depend mainly on mode of deposition, not on stress history; dense materials may have a high strength; clays may have low residual strength discontinuities (e.g. fissures and joints) as for alluvial clays | Original landforms such as buried channels are obscured and the interface with rock may be difficult to determine; severe artesian pressures may exist below tills mantling valley slopes; boulders cause problems with drilling and sampling; pressure meter tests may be useful; properties are controlled by the amount of fine matrix and the presence of clay minerals | Tills generally make good foundation materials, but can be strong when dry and difficult to excavate; boulders cause problems in piling, tunnelling, excavation and filling; drag structures in weak rocks at the base of tills cause errors in rock level estimation, problems with piles, etc.; high permeability layers or lenses may cause water flows into excavations and short-term slope instability; problems with embankment fill as for alluvial clays |
| *Periglacial relict soils (can also be untransported)* (Figures 2.6, 3.2, 3.5, 5.1, 5.3, 5.4) | In temperate climates, relict landscape features that developed under former permafrost conditions may be present generally or locally; former ground freezing is likely to have produced extensive solifluction and colluvial deposits on slopes and to have fractured, brecciated and de-cemented parent soils and rocks; ground often contorted by freeze–thaw features, e.g. cryoturbation and ice wedges | Valley slopes will be affected by strata disturbance and shearing due to valley bulging and cambering; near-vertical fissures and gulls difficult to locate by conventional drilling; relict active layer detachment and solifluction slides are difficult to recognize in landscapes modified under present temperate climatic conditions (only about the last c. 12,000 years in Britain) | Effects of cambering on slopes and valley bulging may cause problems with foundations, excavations, tunnels and groundwater flows; ice-wedge casts may cause local instability in cuttings; potential reactivation of relict periglacial slides formed on slopes as flat as 2–10°; haul road surfaces become slippery for traffic in wet weather |
| *Organic peaty soils (generally untransported)* (Figures 2.1, 3.2, all Parts 4 and 5 figures) | Highly compressible and subject to severe long-term creep; commonly very low unit weight; methane gas may be present; sudden bog-bursts may occur in saturated peat on very gentle slopes; acid near-surface water | Most methods of sampling and testing not suitable for highly organic peaty materials, particularly when fibrous; compressibility usually more important than strength and best measured using large Rowe cells or, for shallow deposits, by large-scale in situ loading tests (plate-bearing, skip test, trial embankment, etc.) | Very large settlements of foundations and embankments requiring digging out and replacing with suitable fill, use of lightweight fills, piling or pre-compression by surcharging and draining; slope stability problems as a result of low passive resistance; non-saturated peat deposits may float when flooded; exposed peats waste and contract on drying; usually difficult to run plant on organic soils and to handle them as spoil; organic acid attack on concrete around ground level |

185

| Soil type and related figures | Typical properties and characteristics | Ground investigation and testing problems | Construction and materials problems |
|---|---|---|---|
| ***Volcanic soils*** *(can also be untransported)* (Figures 1.4, all Part 4 figures) | Properties differ significantly from those of sedimentary soils as a result of the porosity and crushability of silt- and sand-sized particles; in situ moisture content unusually high; increase in stress gives marked reduction in strength, but smaller reduction in compressibility; fine soils often have high plasticity due to smectite or allophane clay minerals, but have higher strength than similar sedimentary soils | Layered and complex deposition including buried former weathered surfaces, palaeosols and thin ash layers weathered to clay that are difficult to identify during fieldwork; investigation by cone testing shows both loose and dense volcanic soils can have similar behaviour to loose quartz sand as a result of crushing of soil grains; compaction curves may have no clear peak | Low particle density makes embankment fills very susceptible to erosion; volcanic soils may soften with compaction and are easily damaged by earth-moving machinery, leading to loss of trafficability and sometimes to soil flow; drying produces non-reversible improvement; addition of quicklime can be effective; weak layers encountered during construction may destabilize cut-slopes |
| ***Untransported*** (formed in situ) soils | | | |
| ***Residual soils*** (Figures 2.2, 2.3, 2.4, 2.5, 3.7, 3.8) | Wide-ranging grading, plasticity, mineralogy and other properties depending primarily on weathering process and the amount of remaining unweathered quartz particles; re-cementation of soils may occur; highly plastic 'black' soils formed in poorly drained areas exhibit large volume changes on wetting and drying; no volume change on wetting/drying of 'red' soils formed in better-drained areas; often have a pronounced structure related to weathering processes; strength and compressibility depend more on structure than on grading, density and mineralogy; bonded structure of soils usually gives high in situ permeability, but yields at a certain stress level; gneisses and granites in stable cratons in the tropics have weathered over tens of millions of years to form residual soils that can be 100 m or more thick | Corestones of unweathered rock within the weathered profile cause problems in drilling; rapidly varying rockhead levels, relict discontinuities from the parent rock, commonly containing iron and manganese oxides, form planes of low drained strength; landslide rupture surfaces are difficult to locate during ground investigations; difficult to obtain 'undisturbed' samples without destroying the soil structure; porous soils, particularly those derived from volcanic rocks, may become sensitive and de-structured at high water contents, giving low undrained strengths on remoulding; the mineralogy of residual soils is often very different from that of non-residual clays; mineralogy and properties can change on drying and can give anomalous results in laboratory testing; aggregated particles of clay and silt may give sand grading until disturbed when they revert to their clay/silt character | Corestones cause problems in piling and may influence open-cut excavation methods; persistent unfavourably orientated weak relict discontinuities can cause cut-slope instability; heavy rain causes severe gullying in erodible soils; structural disturbance and wetting of dry porous soils may substantially reduce their strength and permeability so that they collapse and consolidate rapidly, causing problems in plant operation, cut-slope instability and poor compaction of embankment fill; general properties of fill are similar to those of alluvial clays that have similar grading and mineralogy; shallow earth slides and flows triggered by heavy rainfall on cut-slopes may block access roads and impede construction work; aggregated fine particles may behave as clays on disturbance (e.g. piling, tunnelling) |
| **Man-made fills** | | | |
| ***Heterogeneous made-ground*** (all Parts 4 and 5 figures) | Very variable; may contain toxic and organic materials, materials subject to decay, voids associated with human artefacts, remains of old constructions and sludge lagoons from sewage, agricultural and industrial processes; original topography obscured by subsequent back-fill, building or vegetation; site history is very important; may contain significant amounts of methane | Old non-engineered fills are likely to be heterogeneous and difficult to investigate; original topography obscured by tipping; compressible and weak organic and alluvial soils may be buried locally; usually investigated by trial pits using simple in situ tests, or by boreholes, testing as for organic soils but including tests for chemical leachates; old mineral workings may be back-filled with local overburden and spoil materials that are difficult to differentiate from in situ strata | Infilled quarries are likely to have had steep sides and openwork talus on quarry floor, giving a rapid transition between natural ground and deep-fill material; waste materials such as mine tailings, mine stone and pulverized fuel ash may be useful as engineered fills, particularly where a relatively low density is required, but may contain deleterious chemicals; buried streams and ponds may contain compressible weak alluvial and organic soils |
| ***'Cohesive' clay-fill*** (all Parts 4 and 5 figures) | Bonded structure of parent clay usually destroyed during excavation and fill placing; laminated and varved clays are usually mixed to produce a low permeability fill; degree of saturation is important; low plasticity clays lose significant undrained strength with small increases in water content in wet weather; rutting of wet plastic clay fills under traffic produces shear surfaces that reduce bulk strength | Potential sources of clay-fill are identified during the desk study stage of ground investigation; there may be local knowledge from previous projects; performance of fill largely determined by initial water content, plasticity and compaction characteristics; trafficability, fill layer thickness and stability during construction are controlled by remoulded undrained strength; haul road surfaces become too slippery for traffic in wet weather | Early compaction and profiling of clay seals surface and reduces infiltration and softening by rain; drying prior to placement requires reliable warm dry weather; compaction using smooth drum rollers may create horizontal surfaces of low strength, so that sheeps-foot rollers are preferred. Long-term strength and stability of embankment fills largely governed by drained strength and pore pressures resulting from rainfall infiltration; deep-seated failures may be governed by short-term undrained strength |

# References

Alexander, D. 1999. Natural hazards. In: Alexander, D.E. & Fairbridge, R.W. (eds). *Encyclopedia of Environmental Science.* Kluwer Academic Press, Dordrecht, pp. 42–425.

Allen, D.J., Darling, W.G., Davies, J., Newell, A.J., Gooddy, D.C. & Collins, A.L. 2014. Groundwater conceptual models: implications for evaluating diffuse pollution mitigation measures. *Quarterly Journal of Engineering Geology and Hydrogeology*, 47, 65–80.

American Society of Civil Engineers (ASCE) 1982. *Engineering and Construction in Tropical and Residual Soils.* ASCE, New York, 735 pp.

Atkinson, J.H. 2000. Non-linear soil stiffness in routine design. *Géotechnique*, 50, 487–508.

Baynes, F.J., Fookes, P.G. & Kennedy, J.F. 2005. The total engineering geology approach applied to railways in the Pilbara, Western Australia. *Bulletin of Engineering Geology and the Environment*, 64, 67–94.

Bell, F.G. 2000. *Engineering Properties of Soils and Rocks,* 4th edn. Blackwell Science, Oxford, 482 pp.

Bell, F. & Culshaw, C. 2005. Chalk landscapes. In: Fookes, P.G., Lee, E.M. & Milligan, G. (eds). *Geomorphology for Engineers.* Whittles Publishing, Dunbeath, pp. 729–756.

Blight, G.E. (ed). 1997. *Mechanics of Residual Soils.* Balkema, Rotterdam, 248 pp.

BS 812: 101–106 1984–94. *Testing Aggregates.* British Standards Institution, London (replaced by BS EN 932: 932-1 to 932-6: 1997–2000. *Tests for General Properties of Aggregates*).

BS 5930 1999 (+A2.2010) *Code of Practice for Site Investigations.* British Standards Institution, London, 207 pp.

Charman, J. & Lee, M. 2005. Mountain environments. In: Fookes, P.G., Lee, E.M. & Milligan, G. (eds). *Geomorphology for Engineers.* Whittles Publishing, Dunbeath, pp. 501–533.

Clayton, C.R.I., Matthews, M.C. & Simons, N.E. 1995. *Site Investigation*, 2nd edn. Blackwell Science, Oxford, 584 pp.

Culshaw, M.G. 2005. From concept towards reality: developing the attributed 3D geological model of the shallow sub surface. *Quarterly Journal of Engineering Geology and Hydrogeology*, 38, 231–284.

Dearman, W. R. 1981. General Report, Session I; Engineering Properties of Carbonate Rocks. Symp. on Eng. Geol. Problems of Construction on Soluble Rocks, Aachen/Essen. *Bulletin of the International Association of Engineering Geology*, 24, 3–17.

Derbyshire, E. & Meng, X.M. 1995. The landslide hazard in North China: characteristics and remedial measures at the Jiaoshuwan and Taishanmiao slides in Tian Shui city, Gansu Province. In: McGregor, D.F.M. & Thompson, D.A. (eds). *Geomorphology and Land Management in a Changing Environment.* Wiley, Chichester, pp. 89–104.

Douglas, I. 2005. Hot wetlands. In: Fookes, P.G., Lee, E.M. & Milligan, G. (eds). *Geormorphology for Engineers.* Whittles Publishing, Dunbeath, pp. 473–500.

Duchaufour, P. 1982. *Pedology, Pedogenesis and Classification* (English translation by T.R. Paton). George, Allen and Unwin, London, 448 pp.

EMERCOM 2013. *Annual Report of the Antistikhiya Centre.* Ministry of Emergency Situations, Moscow.

Fookes, P. G. 1988. The geology of carbonate soils and rocks and their engineering characteristics and description. In: Jewell, R. J. & Khorshid, M. S. (eds). *Engineering for Calcareous Sediments.* Balkema, Vol. 2, pp, 789–806.

Fookes, P.G. 1997a. First Glossop Lecture: Geology for Engineers: the geological model, prediction and performance. *Quarterly Journal of Engineering Geology*, 30, 293–424.

Fookes, P.G. (ed). 1997b. *Tropical Residual Soils.* Engineering Group Working Party Revised Report, Geological Society Professional Handbooks. Geological Society, London, 184 pp.

Fookes, P.G., Sweeney, M., Manby, C.N.D. & Martin, R.P. 1985. Geological and geotechnical engineering aspects of low cost roads in mountainous terrain. *Engineering Geology*, 21, 1–152.

Fookes, P.G. and Marsh, A.H. (1981a) Some characteristics of construction materials in the low to moderate metamorphic grade rocks of the Lower Himalayas of East Nepal. 1: Occurrence and geological features. *Proceedings of the Institution of Civil Engineers*, Part 1 70, 123–138.

Fookes, P.G. and Marsh, A.H. (1981b) Some characteristics of construction materials in the low to moderate metamorphic grade rocks of the Lower Himalayas of East Nepal. 2: Engineering characteristics. *Proceedings of the Institution of Civil Engineers*, Part 1, 70, 139–162.

Fookes, P.G., Baynes, F.K. & Hutchinson, J.N. 2000. Total geological history: a model approach to the anticipation, observation and understanding of site conditions. In: *GeoEng 2000, an International Conference on Geotechnical and Geological Engineering*, Vol. 1, pp. 370–460.

Fookes, P.G., Lee, E.M. & Milligan, G. (eds). 2005. *Geomorphology for Engineers.* Whittles Publishing, Dunbeath, 851 pp.

Fookes, P.G., Lee, E.M. & Griffiths, J.S. 2007. *Engineering Geomorphology. Theory and Practice.* Whittles Publishing, Dunbeath, 279 pp.

Francis, P. 1993. *Volcanoes.* Clarendon Press, Oxford, 443 pp.

Grant, K. (ed). 1968. *Proceedings of Study Tour and Symposium on Terrain Evaluation for Engineering.* Division of Applied Geomechanics, CSIRO, Victoria, 101 pp.

Head, K.H. 2006. *Manual of Soil Laboratory Testing*, Vol. 1, 3rd edn. Whittles Publishing, Dunbeath, 424 pp.

Head, K.H. and Epps, R. 2011. *Manual of Soil Laboratory Testing*, Vol. 2, 3rd edn. Whittles Publishing, Dunbeath, 510 pp.

Head, K.H. and Epps, R. 2014. *Manual of Soil Laboratory Testing*, Vol. 3, 3rd edn. Whittles Publishing, Dunbeath, 426 pp.

Hearn, G.J. (ed). 2011. *Slope Engineering for Mountain Roads.* Engineering Geology Special Publication No. 24. Geological Society, London, 301 pp.

Highways Agency, 2007. *Specification for Highways Works.* HMSO, London.

Hutchinson, J.N. 1991. Periglacial and slope processes. In: Forster, A., Culshaw, M.G., Cripps, J.C., Little, J.A. & Moon, C.F. (eds). *Quaternary Engineering Geology.* Engineering Geology Special Publication No. 7. Geological Society, London, pp. 283–331.

International Union of Soil Sciences Working Group 2006. *World Reference Base for Soil Resources.* World Resources Reports No. 103. Food and Agriculture Organisation of the United Nations, Geneva, 145 pp.

Lawrance, C.J. 1972. *Terrain Evaluation in West Malaysia, Part 1.* Report No. TRRL LR 506. Transport and Road Research Laboratory, Crowthorne, 32 pp.

Lawrance, C.J., Byard, R.J. & Beaven, P.J. 1993. *Terrain Evaluation Manual.* Transport Research Laboratory State of the Art Review No. 7. HMSO, London, 285 pp.

Lee, M. & Fookes, P. 2005. Hot drylands. In: Fookes, P.G., Lee, E.M. & Milligan, G. (eds). *Geomorphology for Engineers.* Whittles Publishing, Dunbeath, pp. 419–453.

McFarlane, M.J. 1983. Laterites. In: Goudie, A.S. & Pye, K. (eds). *Chemical Sediments and Geomorphology.* Academic Press, London, pp. 7–58.

Miller, G.H., Lehman, S.J., Refsnider, K.A., Southon, J.R. & Zhong, Y. 2013. Unprecedented recent summer warmth in Arctic Canada. *Geophysical Research Letters*, 40, 5745–5751.

Morgenstern, N.R. & Cruden, D.M. 1997. Description and classification of geotechnical complexities. In: *Proceedings of the International Symposium on the Geotechnics of Structurally Complex Formations, 2.* Associazione Geotechnica Italiana, Rome, pp. 195–204.

Mortimore, R. N., Wood, C. J. & Gakkois, R. W. 2001. British Upper Cretaceous Stratigraphy. *Geological Conservation Review Series No. 23.* Nature Conservancy Committee, Peterborough, 558

Nicholson, D., Tse, C.M. & Penny, C. 1999. *The Observational Method in Ground Engineering: Principles and Applications.* CIRIA Report R185, CIRIA, London, 214 pp.

Norbury, D. 2010. *Soil and Rock Description in Engineering Practice.* Whittles Publishing, Dunbeath, 288 pp (contains helpful list of latest relevant British Standards and Euro Codes).

Parry, S., Baynes, F.J., Culshaw, M.G., Eggers, M., Keaton, J.R., Lentfer, K., Novotny, J. & Paul, D. 2014. Engineering geology models: an introduction. IAEG Commission 25. *Bulletin of Engineering Geology and the Environment*, 73, 689–706.

Perry, J. & West, G. 1996. *Preliminary Sources of Information for Site Investigation in Britain. (Revision of TRL Report LR403).* TRL Project Report 192. Transport Research Laboratory, Crowthorne, 47 pp.

Pettifer, G.S. & Fookes, P.G. 1984. A revision of the graphical method of accessing the excavatability of rock. *Quarterly Journal of Engineering Geology*, 27, 145–164.

Prance, G.T. 2002. Tropical forests. In: Mooney, H.A. & Canadell, J.G. (eds). *The Encyclopaedia of Global Environmental Change: The Earth System – Biological and Ecological Dimensions of Global Environmental Change*, Vol. 2. Wiley, Chichester, pp. 582–586.

Rawson, P.F., Allen, P. & Gale, A. 2001. The Chalk Group – a revised lithostratigraphy. *Geoscientist*, 11, 1, 21.

Ria Novosti 2013. Arctic temperatures to grow twice as fast – Russian ministry (online). Available from: http://en.ria.ru/russia/archive/20130320.

Richardson, S. D. & Reynolds, J. M. 2000. An overview of glacial hazards in the Himalayas. *Quaternary International*, 65/66, 31–47.

Rowe, P.W. 1972. The relevance of soil fabric to site investigation practice. *Géotechnique*, 22, 192–300.

Selley, R.C. 1996. *Ancient Sedimentary Environments and their Sub-surface Diagnosis*, 4th edn. Chapman and Hall, London, 300 pp.

Sidorchuk, A. & Grigor'ev, V. 1998. Soil erosion on the Yamal Peninsula (Russian Arctic) due to gas field exploitation. *Advances in GeoEcology*, 31, 805–811.

Simons, N., Menzies, B. & Matthews, M. 2002. *A Short Course in Geotechnical Site Investigation.* Thomas Telford, London, 353 pp.

Smith, M.R. (ed). 1999. *Stone: Building Stone, Rock Fill and Armourstone in Construction.* Engineering Geology Special Publication No. 16. Geological Society, London, 478 pp.

Smith, M.R. & Collis, L. (eds). 2001. *Aggregates: Sand, Gravel and Crushed Rock for Construction Purposes*, 3rd edn. Revised by Fookes, P.G., Lay, J., Sims, I., Smith, M.R. & West, G (eds). Engineering Geology Special Publication No. 17. Geological Society, London, 339 pp.

Stoddart, D. 1969. Climatic geomorphology: review and re-assessment. *Progress in Geography*, 1, 159–222.

Strakhov, N.M. 1967. *Principles of Lithogenesis*, Vol. I. Oliver and Boyd, Edinburgh, 245 pp.

Sweeney, M. (ed). 2004. *Terrain and Geohazard Challenges facing Onshore Oil and Gas Pipelines.* Thomas Telford. London. 735.

Thomas, M.F. 2005. Savanna. In: Fookes, P.G., Lee, E.M. & Milligan, G. (eds). *Geomorphology for Engineers.* Whittles Publishing, Dunbeath, pp. 454–473.

Trenter, N.A. 1999. *Engineering in Glacial Tills.* CIRIA Report C504. CIRIA, London, 259 pp.

Tricart, J. 1957. Application du concept de zonalité à la géomorphologie (in French). *Tijdschrift van het koninklijk Nederlandsch Aardrijkskundig Genootschap*, 74, 422–434.

Tricart, J. & Cailleux, A. 1965. *Introduction á la Géomorphologie Climatique* (in French). SEDES, Paris, 306 pp.

Ulusay, R. & Hudson, J.A. (eds). 2007. *The Complete ISRM Suggested Methods for Rock Characterisation, Testing and Monitoring: 1974–2006.* International Society for Rock Mechanics, Lisbon, 628 pp.

Vaughan, P.R. 1994. Thirty-fourth Rankine Lecture: Assumption, prediction and reality in geotechnical engineering. *Géotechnique*, 44, 571–609.

Walker, H.J. 2005. Periglacial forms and processes. In: Fookes, P.G., Lee, E.M. & Milligan, G. (eds). *Geomorphology for Engineers.* Whittles Publishing, Dunbeath, pp. 377–399.

Walker, M.J. (ed). 2012. *Hot Deserts: Engineering, Geology and Geomorphology.* Engineering Group Working Party Report.

Geological Society Special Publication No. 25. Geological Society, London, 425 pp.

Waltham, A.C. & Fookes, P.G. 2003. Engineering classification of karst ground conditions. *Quarterly Journal of Engineering Geology and Hydrogeology*, 36, 101–118.

Waltham, T. 2009. *Foundations of Engineering Geology,* 3rd edn. Spon Press, London, 98 pp.

## Bibliography

In addition to the textbooks listed in the references, the following relevant books used by the authors in their professional work are given for background reading. These are subdivided into two groups (A and B). The list is not necessarily comprehensive, but is considered to be helpful for readers of this book, either specifically or as general background knowledge.

\* Useful introduction; \*\* more specialized or detailed; \*\*\* relates to parts of this book.

### Group A: Introductory geology and geomorphology textbooks

\*\*Ballantyne, C.K. & Harris, C. 1994. *The Periglaciation of Great Britain.* Cambridge University Press, Cambridge, 330 pp.

\* Blyth, F.G.H. & De Freitas, M.H. 1984. *A Geology for Engineers,* 7th edn. Edward Arnold, London, 325 pp.

\* Chernicoff, S. 1999. *Geology.* Houghton Mifflin, Boston, 640 pp.

\*\*Collinson, J.D. & Thompson, D.B. 1989. *Sedimentary Structures,* 2nd edn. Unwin Hyman, London, 207 pp.

\*\*Cooke, R.U., Brunsden, D., Doornkamp, J.C. & Jones, D.K.C. 1982. *Urban Geomorphology in Drylands.* Oxford University Press, Oxford, 324 pp.

\*\*Dott Jr. R.H. & Batten, R.L. 1988. *Evolution of the Earth,* 4th edn. McGraw-Hill, New York, 120 pp.

\*\*Duff, P.McL.D. & Smith, A.J. 1992. *Geology of England and Wales.* The Geological Society, London, 651 pp.

\* Hamblin, W.K. & Christiansen, E.H. 1995. *Earth's Dynamic Systems,* 7th edn. Prentice-Hall, New Jersey, 710 pp.

\* Huggett, R.J. 2007. *Fundamentals of Geomorphology,* 2nd edn. Routledge, Abingdon, 458 pp.

\* Kious, W.J. & Tilling, R.I. 1996. *This Dynamic Earth: the Story of Plate Tectonics.* US Geological Survey, Denver, CO, 77 pp.

\*\*Mason, R. 1990. *Petrology of the Metamorphic Rocks,* 2nd edn. Unwin Hyman, London, 230 pp.

\*\*\*Pettijohn, F.J. 1983. *Sedimentary Rocks,* 3rd edn. Harper Collins, New York, 628 pp.

\* Press, F. & Siever, R. 2001. *Understanding Earth,* 3rd edn. W.H. Freeman, New York, 624 pp.

\*\*\*Summerfield, M.A. 1991. *Global Geomorphology.* Longman, Harlow, 537 pp.

\* Tarbuck, E.J. & Lutgens, F.K. 1996. *Earth. An Introduction to Physical Geology.* Prentice Hall, NJ, 605 pp.

\* Toghill, P. 2000. *The Geology of Britain. An Introduction.* Swan Hill Press, Shrewsbury, 192 pp.

\* Tucker, M.E. 2001. *Sedimentary Petrology: an Introduction to the Origin of Sedimentary Rocks,* 3rd edn. Blackwell Science, Oxford, 262 pp.

### Group B: Engineering geology, engineering geomorphology, geotechnical and site investigation textbooks

\*\*Anon. 1995. The description and classification of weathered rocks for engineering purposes. Engineering Group Working Party Report. Quarterly Journal of Engineering Geology and Hydrogeology, 28, 207–242.

\*\*Burland, J., Chapman, T., Skinner, H. & Brown, M. (eds). 2012. *ICE Manual of Geotechnical Engineering, Volume 1: Geotechnical Engineering Principles, Problematic Soils and Site Investigation.* ICE Publishing, London, 650pp.

\*\*Burland, J., Chapman, T., Skinner, H. & Brown, M. (eds). 2012. *ICE Manual of Geotechnical Engineering, Volume 2: Geotechnical Design, Construction and Verification.* ICE Publishing, London, 750pp.

\* Goudie, A.S. & Brunsden, D. 1994. *The Environment of the British Isles. An Atlas.* Clarendon Press, Oxford, 184 pp.

\*\*Griffiths, J.S. (ed). 2001. *Land Surface Evaluation for Engineering Practice.* Engineering Geology Special Publication No. 18. Geological Society, London, 248 pp.

\*\*Hearn, G.J. (ed). 2011. *Slope Engineering for Mountain Roads.* Engineering Geology Special Publication No. 24. Geological Society, London, 301 pp.

\* McClay, K.R. 1991. *The Mapping of Geological Structures.* Geological Society of London Professional Handbook Series. Wiley, Chichester, 168 pp.

\* Price, D.G. 2009. *Engineering Geology. Principles and Practice.* Edited and compiled by M.H. De Freitas. Springer, Berlin, 450 pp.

\* Radford, T.A. (ed). 2012. *Earthworks in Europe.* Engineering Geology Special Publication No. 26. Geological Society, London, 185 pp.

\* West, G. 1991. *The Field Description of Engineering Soils and Rocks.* Geological Society of London Professional Handbook Series. Open University Press, Milton Keynes, 129 pp.

## Locations of photographs

*Photographs on each spread are numbered from left to right, and then from top to bottom, first on the left page, and then on the right page; as in the example shown for section 1.6.*

**1-1 Morphoclimatic zones**
1. Torrel Land, Spitsbergen, Svalbard, Norway.
2. Ogilvie Mountains, Yukon, Canada.
3. Trent Valley, Nottingham, UK.
4. Yosemite Valley, California, USA.
5. Trans-Alaska Oil Pipeline, Delta Junction, Alaska, USA.
6. Iya Valley, Niyodo, Shikoku, Japan.
7. Mamallapuram, Tamil Nadu, India.
8. Death Valley, California, USA.
9. Pilar, Bicol, southern Luzon, Philippines.
10. Amdo, Tibet, China.
11. Island in the Sky, Canyonlands, Utah, USA.
12. Kikori Valley, Papua New Guinea.

**1-3 Crustal plate, volcanoes and earthquakes**
1. Himalayas from Pang La, Tingri, Tibet.
2. Grjotagja, Myvatn, northeastern Iceland.
3. Breiddalur, eastern Iceland.
4. Aleutian Islands, Alaska, USA.
5. Point Reyes Station, Marin, California, USA.
6. Eqi, Disko Bay, western Greenland.
7. Hammersley Gorge, Pilbara, Western Australia.
8. Karakoram, Kashmir, Pakistan.
9. Loch Tulla, Rannoch Moor, Scotland, UK.
10. Point Reyes, San Francisco, California, USA.
11. Calaveras Fault, 6th Street, Hollister, California, USA.
12. Great Sandy Desert, Broome, Western Australia.

**1-4 Igneous rock associations**
1. Whin Sill, Hadrian's Wall, Northumbria, UK.
2. Shiprock, Farmington, New Mexico, USA.

3. Wadi Yutum, Aqaba, Dead Sea Rift, Jordan.
4. Buddoso, Nuoro, Sardinia, Italy.
5. Karimsky volcano, Kamchtaka, Russia.
6. Mount Popa, Bagan, Myanmar.
7. Soufrière Hills, Montserrat, West Indies.
8. El Misti, Arequipa, Peru.
9. Pompeii, Naples, Italy.
10. Avacha and Koryaksky, Kamchatka, Russia.
11. Pu'u O'o, Kilauea, Kalapana, Hawaii, USA.
12. Aldeyarfoss, Iceland.
13. Kailua Kona, Big Island, Hawaii, USA
14. Saddle, Mauna Kea, Hawaii, USA

**1-5 Metamorphic rock associations**
1. North Uist, Outer Hebrides, Scotland, UK.
2. Hammerfest, Finnmark, Norway.
3. Nantlle Quarry, Snowdonia, Wales, UK.
4. Bethesda Quarry, Snowdonia, Wales, UK.
5. Mount Rushmore, Keystone, South Dakota, USA.
6. Monte Corchia, Carrara, Tuscany, Italy.
7. Salisbury Crag, Edinburgh, Scotland, UK.
8. El Tatio, Andes, Calama, Norte Grande, Chile.
9. Hemerdon, Dartmoor, Devon, UK.
10. Darcha, Chenab Valley, Himachal Pradesh, India.

**1-6 Clastic sediments**

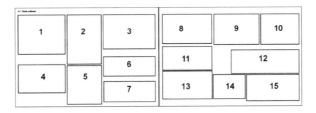

1. Alum Bay, Isle of Wight, UK.
2. Chenab Valley, Lahaul, Himachal Pradesh, India.
3. Specimens from UK.
4. Skipsea, Holderness, East Yorkshire, UK.

5. Lulworth Cove, Dorset, UK.
6. Kvaenangen Sorfjord, Troms, Norway.
7. Wailua, Kauai, Hawaii, USA
8. Mawddach Estuary, Snowdonia, Wales, UK.
9. River Axe, Seaton, South Devon, UK.
10. Vlamingh Head, Cape Range, Western Australia.
11. Bognor Regis, West Sussex, UK.
12. Hatteras Island, North Carolina, USA.
13. Taormina, Sicily, Italy.
14. North Uist, Outer Hebrides, Scotland, UK.
15. Okefenokee Swamp, Georgia and Florida, USA.

**1-8 Clastic sedimentary rocks**
1. Capitol Reef, Utah, USA.
2. Healy Mining District, Denali, Alaska, USA.
3. Carperby flagstone mine, Wensleydale, Yorkshire, UK.
4. Stanage Edge, Peak District, Derbyshire, UK.
5. Monticello, Utah, USA.
6. Castle Rock, Nottingham, UK.
7. Zabriske Point, Death Valley, California, USA.
8. Bingling Si, Lanzhou, Gansu, China.
9. The Narrows, Zion Canyon, Utah, USA.
10, Brimham Rocks, Nidderdale, North Yorkshire, UK.
11. Torres del Paine, Patagonia, Chile.
12. Painted Desert, Holbrook, Arizona, USA.
13. Torres del Paine, Patagonia, Chile.

**1-9 Carbonate rocks**
1. South Beach, Mana Island, Viti Levu, Fiji.
2. Aqaba, Red Sea, Jordan.
3. Hundred Islands, Lingayen Gulf, Luzon, Philippines.
4. Mamanuca Islands, Viti Levu, Fiji.
5. Scaraster, South Harris, Outer Hebrides, Scotland, UK.
6. Shell Beach, Shark Bay, Western Australia.
7. Muschelkalk Limestone, Baden Württemberg, Germany.
8. Seven Sisters, South Downs, East Sussex, UK.

9. Gunung Sewu, Java, Indonesia.

10. Salar de Uyuni, Potosi, Bolivia.

11. Lyme Regis, Dorset, UK.

12. High Tor, Matlock Bath, Derbyshire, UK.

13. Gordale Scar, Malham, Yorkshire Dales, UK.

## 1-10 Rock mass discontinuities

1. Castlemartin, Dyfed, Wales, UK.

2. Stair Hole, Lulworth Cove, Purbeck, Dorset, UK.

3. Millook Haven, Bude, Cornwall, UK.

4. Pang La, Tingri, Tibet.

5. Blue Mountain, Dinosaur, Colorado, USA.

6. Setwicks Bay, Flamborough Head, East Yorkshire, UK.

7. Pecos Valley, Carlsbad, New Mexico, USA.

8. Kemmerer, Utah, USA.

9 and 10. Tsepelovon, Tymfi Oros, northern Greece.

11. Foredale Quarry, Horton, Yorkshire Dales, UK.

12. Stocksbridge, South Yorkshire, UK.

13. Semail Pass, Nizwa, Oman.

14. Long Scars, Ingleborough, Yorkshire Dales, UK.

15. Jabal Qasyun, Damascus, Syria.

## 2-1 Soils and soil models

1. Cache Creek, Colusa, California, USA.

2. Lipari, Aeolian Islands, Sicily, Italy.

3. Trent Valley, Nottinghamshire, UK.

4. Mexican Hat, Utah, USA.

5. Monteverde, Santa Elena, Costa Rica.

6. Moa, eastern Holguin, Cuba.

7. Ross of Mull, Inner Hebrides, Scotland, UK

8. Skeleton Coast, Namibia.

9. Arbol de Piedra, Siloli Desert, Potosi, Bolivia.

10. Kinder Scout, Peak District, Derbyshire, UK.

11. Barnatra, County Mayo, Ireland.

12. Tarif, Abu Dhabi, United Arab Emirates.

13. Stockertown, Pennsylvania, USA.

## 2-3 Weathering processes

1. Lüderitz, Namibia.

2. Shiprock, Farmington, New Mexico, USA.

3. Limestone Gorge, Bullita, Northern Territory, Australia.

4. Half Dome, Yosemite, California, USA.

5. Kettlewell, Yorkshire Dales, UK.

6. Tafraoute, Anti Atlas, Morocco.

7. Musandam Peninsula, Oman.

8. Waimea Canyon, Kauai, Hawaii, USA.

9. Beggar's Gate, Ingleborough, Yorkshire Dales, UK.

10. Chenab Valley, Lahaul, Himachal Pradesh, India.

11. Zillertal, Tyrol, Austria.

12. Nottingham, UK.

## 2-4 Granite weathering

1. Hammersley Ranges, Pilbara, Western Australia.

2. Wiluna, Western Australia.

3. Fitzroy Crossing, Western Australia.

4. Kowloon, Hong Kong.

5. Batad, Banaue, Luzon, Philippines.

6. Salt River Canyon, Arizona, USA.

7. Alabama Hills and Mount Whitney, California, USA.

8. Devil's Marbles, Northern Territory, Australia.

9. Ipoh, Perak, Malaya, Malaysia.

10. Stanton, Peak District, Derbyshire, UK.

11. Giant's Causeway, County Antrim, Northern Ireland, UK.

12. Kakaha, Kauai, Hawaii, USA.

## 2-5 Slopes and valley sides

1. Capitol Reef, Torrey, Utah, USA.

2. Sarchu, Zanskar, Ladakh, India.

3. Wastwater, Lake District, Cumbria, UK.

4. Jabal at-Tar, Palmyra, Syria.

5. Adventdale, Spitsbergen, Svalbard, Norway.

6. El Calafate, Patagonia, Argentina.

7. Wadi al Hasa, Tafila, Jordan.

8. Whitecliffe Bay, Isle of Wight, UK.

9. Panamint Valley, California, USA.

10. Khazzan Desert, Oman.

11. Iuiu, Minas Gerais, Brazil.

## 3-1 Glaciated environments

1. Tingkya Himalayas, Tibet.

2. Trollstigen, Andalsnes, western Norway.

3. East Greenland coastal mountains.

4. Alta, northern Norway.

5. Meade Glacier, Juneau Icefield, Alaska, USA.

6. Gryllefjord, Senja, northern, Norway.

7. Lysefjord, Stavanger, Norway.

8. Rongbuk Valley, southern Tibet.

9. Laguna Torre, Patagonia, Argentina.

10. Karnali Valley, western Nepal.

11 and 12. Carstairs, Clyde Valley, Scotland, UK.

13. Ribblehead, Yorkshire Dales, UK.

14. Norber, Ingleborough, Yorkshire Dales, UK.

15. Val Veni, Monte Bianco, Italy.

16. Miage Glacier, Monte Bianco, Italy.

17. Alport, Peak District, Derbyshire, UK.

## 3-2 Periglacial environments

1. Ogilvie Mountains, Yukon, Canada.

2. Dempster Highway, Yukon, Canada.

3. White Pass, Skagway, Alaska, USA.

4. Columbia Icefield, Rocky Mountains, Alberta, Canada.

5. Pang La, Himalayas, Tibet.

6. Nara Lagna Pass, Himalayas, western Nepal.

7. Rainbow Mountain, Glenallen, Alaska, USA.

8. Tangle Lakes, Denali Highway, Alaska, USA.

9. Disko Island, western Greenland.

10. Kapp Linné, Spitsbergen, Svalbard, Norway.

11. Ogilvie Mountains, Yukon, Canada.

12. Longyearbyen, Spitsbergen, Svalbard, Norway.

13 and 14. Prudhoe Bay, northern Alaska.

## 3-3 Temperate fluvial environments

1. Kikori Valley, Papua New Guinea.
2. Carrock Fell, Cumbria, UK.
3. Southern Alberta, Canada.
4. Surprise Creek, Litchfield, Northern Territory, Australia.
5. Langtang Valley, Nepal.
6. Birkdale, Swaledale, Yorkshire Dales, UK.
7. Whaw, Arkengarthdale, Yorkshire Dales, UK.
8. Kylesku, Sutherland, Scotland, UK.
9. Hoveringham, Trent Valley, Nottinghamshire, UK.
10. Lagan River, Islay, Inner Hebrides, Scotland, UK.
11. Na'ur, East Bank Plateau, Jordan.

## 3-4 Temperate Mediterranean

1. Peyrepertuse, Aude, France.
2. Vale do Zêzere, Sierra da Estrela, Portugal.
3. Irati Valley, Pamplona, Navarra, Spain.
4. Belo Horizonte, Minas Gerais, Brazil.
5. Irati Valley, Pamplona, Navarra, Spain.
6. Ürgüp, Cappadocia, Nevsehir, Turkey.
7. Kashgar, Xinjiang, China.
8. Métlaoui, Gafsa, Tunisia.
9. Tarn Gorge, Causse Mejean, Lozère, France.
10. Laurac, Aude, France.
11. Imranli, Sivas, Turkey.
12. River Trent, Nottingham, UK.

## 3-5 Relict periglacial features in Britain

1. Thieves Moss, Ingleborough, Yorkshire Dales, UK.
2. Quirang, Isle of Skye, Scotland, UK.
3. Walton Common, Swaffham, Norfolk, UK.
4. Bracadale, Isle of Skye, Scotland, UK.
5. Hawkswick Clowder, Yorkshire Dales, UK.
6. Houndstor, Dartmoor, Devon, UK.
7. Bruichladdich, Islay, Scotland, UK.
8. Lanzhou, Gansu, China.
9. Long Dale, Peak District, Derbyshire, UK.

10. Winspit, Purbeck, Dorset, UK.
11. Pleasley, Mansfield, Nottinghamshire, UK.
12. Black Rock, Brighton, Sussex, UK.
13. Castle Lime Works, South Mimms, Hertfordshire, UK.
14. White Horse Hill, Marlborough Downs, Wiltshire, UK.
15. Folkestone Warren, Kent, UK.
16. Devil's Punchbowl, Brecklands, Norfolk, UK.

## 3-6 Hot desert environments

1. Dakhla basin, Western Desert, Egypt.
2. Antelope Canyon, Page, Arizona, USA.
3. Sossusvlei, Namib Desert, Namibia.
4. Jebel al Batra, southern Jordan.
5. Simpson Desert, Northern Territory, Australia.
6. Dakhla basin, Western Desert, Egypt.
7. Death Valley, California, USA.
8 and 9. Kharga basin, Western Desert, Egypt.
10. Panamint Valley, California, USA.
11 and 12. Lake Assal, Djibouti.
13. Great Sandy Desert, Western Australia.
14. Nizwa, Oman.
15. Todra Gorge, Tinerhir, Morroco.

## 3-8 Hot wet tropical environments

1. Boca San Carlos, Rio San Juan, Costa Rica.
2. Niah, Miri, Sarawak, Malaysia.
3. Cape Tribulation, Daintree, Queensland, Australia.
4. Melinau River, Gunung Mulu, Sarawak, Malyasia.
5. Nausori Highlands, Viti Levu, Fiji.
6. Kikori Valley, Papua New Guinea.
7. Thibodaux, Louisiana, USA.
8. Babeldaob, Palau, Micronesia.
9. Clarendon Hills, Jamaica.
10. Yangshuo, Guangxi, China.
11. Aberdeen, Hong Kong.
12. Banaue, northern Luzon, Philippines.
13. The Pinnacles, Gunung Mulu, Sarawak, Malaysia.

## 3-9 Mountain environments

1 and 2. Trisuli, Langtang Himalaya, Nepal.
3. Halsema Highway, Cordillera, Luzon, Philippines.
4. More Basin, Zanskar, Ladakh, Indian Himalayas.
5. Min Jiang valley, northern Sichuan, China.
6. Spiti Valley, Himachel Pradesh, Indian Himalayas.
7. Karakoram Highway, northern Pakistan.
8. Halsema Highway, Cordillera, Luzon, Philippines.
9. Lamayuru, Zanskar Range, Ladakh, India.
10. Kali Gandaki Valley, Nepal Himalayas.
11. Tafjord (1934 event), western Norway.
12. Taroko Gorge, Taiwan.
13. Quake Lake (1959 event), Madison, Montana, USA.
14. Elda, Alicante, Spain.
15. Hunza Valley and Rakaposi, Karakoram, Pakistan.

## 4-3 Walkover survey

1. Cray, Wharfedale, Yorkshire Dales, UK.
2. Colville, Washington, USA.
3. Kindia, Guinea.
4. South Downs, Fulking, West Sussex, UK.
5. Ethiopian Highlands, Ethiopia.
6. Cabin Creek, Madison River Canyon, Montana, USA.
7. Inkersall Green, Derbyshire, UK.
8. Bonsall Moor, Matlock, Derbyshire, UK.
9. Mam Nick, Edale, Derbyshire Peak District, UK.
10. Widdale, Yorkshire Dales, UK.
11. Cressbrook Dale, Derbyshire Peak District, UK.
12. Kikori Valley, Papua New Guinea.
13. Knockshinnock Colliery (1950 event), Ayrshire, UK.

## 4-7 Ground investigation in gently dipping strata

1. Temple Newsam, Leeds, Yorkshire, UK.
2. Chesterfield, Derbyshire, UK.
3. Llanwrst, Clwyd, Wales, UK.
4. Ashop Valley, Derbyshire Peak District, UK.
5. Walsall, West Midlands, UK.

6. Ilkeston, Derbyshire, UK.

7. Mansfield, Nottinghamshire, UK.

8 and 9. Nottingham, UK.

10 and 11. Jecheon, Korea.

12. Khazzan Desert, Oman.

13. Trinidad, Caribbean.

## 5-2 Complex structures GI

1. Eglwyseg Mountain, Clwyd, Wales, UK.

2. Sundalsvatnet, Rogaland, southern Norway.

3. Grand Canyon below Mohave Point, Arizona, USA.

4. Turtle Mountain (1903 event), Frank, Alberta, Canada.

5. Quirang, Isle of Skye, Scotland, UK.

6. Glen Roy, Highland, Scotland, UK.

7. Skarvberget, Porsangerfjorden, Finnmark, Norway.

8. Lauterbrunnen, Bernese Oberland, Switzerland.

9. Xizi valley, Gilazi, Azerbaijan.

10. Sarchu, Zanskar Range, Ladakh, India.

## 5-3 North Wales coast road

All in Clwyd and Gwynedd, Wales, UK.

## 5-5 Karst

1. Balikesir, western Turkey.

2. Shuicheng, Guizhou, China.

3. Huntsville, Alabama, USA.

4. Blacklion, Cavan, Ireland.

5. Marble Pot, Ingleborough, Yorkshire Dales, UK.

6. Yangshuo, Guilin, Guangxi, China.

7. Sinkhole Plain, Bowling Green, Kentucky, USA.

8. Dunn's River Falls, Ocho Rios, Jamaica.

9. Chocolate Hills, Bohol, Philippines.

10. Ogof Agen Allwedd, Llangattock, Powys, Wales, UK.

11. Popovo Polje,  Trebinje, Bosnia and Herzegovina.

## 5-6 Karst construction

1. Ure Terrace, Ripon, North Yorkshire, UK.

2. Divača, Postojna, Slovenia.

3. Kutaisi, Georgia.

4. Brookwood, Stockertown, Pennsylvania, USA.

5. Fontwell, West Sussex, UK.

6. Remouchamps Viaduct, Liége, Belgium.

7. Guilin, Guangxi, China.

8. County Courthouse, Huntsville, Alabama, USA.

9. Northwich, Cheshire, UK.

10. Sagada, northern Luzon, Philippines.

11. Natural Tunnel, Duffield, Virginia, USA.

## 5-7 Quarry workings

1. Alport, Derbyshire Peak District, UK.

2. Bingham Canyon, Salt Lake City, Utah, USA.

3. Moss Rake, Castleton, Derbyshire Peak District, UK.

4. Centurion, Johannesburg, South Africa.

5. Black Thunder Mine, Gillette, Wyoming, USA.

6. Stockertown, Pennsylvania, USA.

7. Hucklow Edge, Derbyshire Peak District, UK.

8. Bingham Canyon, Salt Lake City, Utah, USA.

9. Shap Pink Quarry, Cumbria, UK.

10. Rock of Ages Quarry, Barre, Vermont, USA.

11. Mission Copper Mine, Tucson, Arizona, USA.

12. Maltby, South Yorkshire, UK.

13. Winspit Mines, Purbeck, Dorset, UK.

14. Buddosso, Nuoro, Sardinia, Italy.

## 5-8 Quarry products

1. Mappleton (stone from Norway), Humberside, UK.

2. Empingham Reservoir, Rutland Water, Rutland, UK.

3 and 6. Hemington, Trent Valley, Nottinghamshire, UK.

4 and 5. Ingleton, Yorkshire Dales, UK.

7. Langton Matravers, Purbeck, Dorset, UK.

8. Plymouth, Devon, UK.

9. Jakarta, Java, Indonesia.

10. Cal Orcko, Sucre, Bolivia.

11. Shell Road, Mississippi Delta, Louisiana, USA.

12. New Orleans, Louisiana, USA.

13. Loscoe (1986 event), Derbyshire, UK.

## 5-9 Arctic environmental problems

1. Atigun Pass, Brooks Range, Alaska, USA.

2. Kapp Linné, Spitsbergen, Svalbard, Norway.

3. Ibyuk Pingo, Tuktoyaktuk, Northwest Territories, Canada.

4. Thjorsa valley, central Iceland.

5. Ice cellar, Tuktoyaktuk, Northwest Territories, Canada.

6. Fox Tunnel, Fairbanks, Alaska, USA.

7. Djupivogur, southeast Iceland.

8. Arctic Plain, Prudhoe Bay, Alaska, USA.

9. Tuktoyaktuk, Northwest Territories, Canada.

10. Prudhoe Bay, Alaska, USA.

11. Inuvik, Northwest Territories, Canada.

12. Dawson, Yukon, Canada.

13. Delta, Alaska, USA.

14. Inuvik, Northwest Territories, Canada.

15. Dachnye, Mutnovsky, Kamchatka, Russia.

16. Dalton Highway, Alaska, USA.

*Photograph credits:*

*1.3.7 by Laurance Donnelly;*

*2.5.10, 4.7.12, 5.8.8 and 5.8.9 by Peter Fookes;*

*4.3.3, 4.3.5 and  4.7.13 by Geoff Pettifer;*

*5.8.4 by Robin Gillespie.*

*All others by Tony Waltham.*

# Index